Rebooting the Herman & Chomsky Propaganda Model in the Twenty-First Century

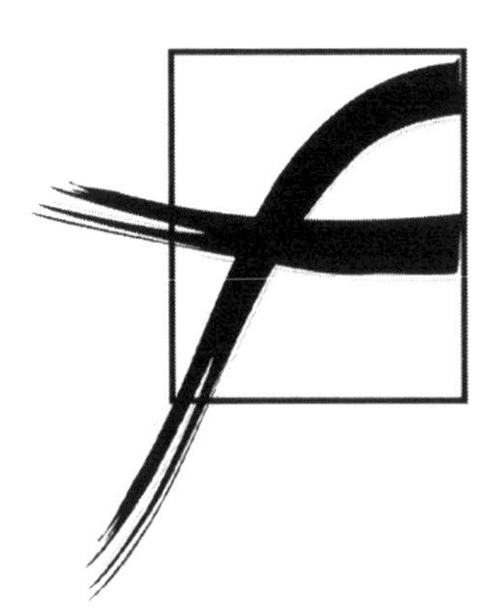

Intersections in Communications and Culture

Global Approaches and Transdisciplinary Perspectives

Cameron McCarthy and Angharad N. Valdivia
General Editors

Vol. 30

The Intersections in Communications and Culture series is part of the Peter Lang Media and Communication list.
Every volume is peer reviewed and meets the highest quality standards for content and production.

PETER LANG
New York • Washington, D.C./Baltimore • Bern
Frankfurt • Berlin • Brussels • Vienna • Oxford

Brian Michael Goss

Rebooting the Herman & Chomsky Propaganda Model in the Twenty-First Century

PETER LANG
New York • Washington, D.C./Baltimore • Bern
Frankfurt • Berlin • Brussels • Vienna • Oxford

Library of Congress Cataloging-in-Publication Data

Goss, Brian Michael.
Rebooting the Herman & Chomsky propaganda model
in the twenty-first century / Brian Michael Goss.
pages cm — (Intersections in communications and culture; vol. 30)
Includes bibliographical references and index.
1. Mass media—Objectivity. 2. Press and propaganda.
3. Propaganda. I. Title.
P96.O242G67 302.23'0973—dc23 2013011870
ISBN 978-1-4331-1621-6 (hardcover)
ISBN 978-1-4331-1620-9 (paperback)
ISBN 978-1-4539-1133-4 (e-book)
ISSN 1528-610X

Bibliographic information published by **Die Deutsche Nationalbibliothek**.
Die Deutsche Nationalbibliothek lists this publication in the "Deutsche
Nationalbibliografie"; detailed bibliographic data is available
on the Internet at http://dnb.d-nb.de/.

The paper in this book meets the guidelines for permanence and durability
of the Committee on Production Guidelines for Book Longevity
of the Council of Library Resources.

29 Broadway, 18th floor, New York, NY 10006
www.peterlang.com

Printed in the United States of America

CONTENTS

ACKNOWLEDGMENTS

In composing this text between May 2011 and December 2012, I never recall having had to give myself a "pep talk" in order to get to go to the work bench and hammer out readings and prose. In feeling this abiding commitment to the topic, I am also indebted to people who helped bring this project into being. Foremost among them are Cameron McCarthy, the series co-editor, and Mary Savigar, the acquisitions editor at Peter Lang who encouraged and contracted the book. The administration at Saint Louis University-Madrid, Spain endowed me with the release time during 2012 that enabled the text to be completed in a timely and, I hope, thorough fashion. For this, I extend thanks to Director Paul Vita, Division Head Laurie Mazzuca and the Faculty Professional Development Advisory Committee for arranging the release time.

My students' questions and comments in classes also helped this project to gestate, most notably in the three editions of Communication 347 ("Political Communication") during the autumn semesters of 2010, 2011, and 2012 when the book was in various stages of being conceptualized and composed. Abigail Schultz, in particular, showed lively undergraduate wonderment at the book's progress. People who read and commented on the proposal and chapters at various moments during its trajectory toward publication include Jack Z. Bratich, Ergin Bulut, Joan Pedro Carañana, Clifford G. Christians, James Curran, Cameron McCarthy, John C. Nerone, Pilar Novoa Salvador, Erika Polson, and Francisco Seoane Pérez. Daniel Chornet and T. Ryan Day engineered a Skype presentation of this work to a seminar at University of Illinois at Urbana-Champaign, while Christopher Chávez and Mary Gould did so for a brown-bag presentation for Saint Louis University's Missouri Campus. Paul Zinder and Adriano Petrucci facilitated my use of office space to write at American University of Rome in March 2012. During the post-writing typesetting process, Sophie Appel of Peter Lang came through in the clutch.

It may not have required an entire village to compose this volume, but I am deeply grateful to the people who took an interest and offered their expertise along the way. Nevertheless, any flaws that endure in the text are my own creation.

INTRODUCTION
News Media Is Not a Sofa or a Plate of Paella

Mass media in general, and news media in particular, are peculiar industries. Part of their specialness resides in being texts that even millions of people can use at the same time without any diminishment in quality—in contrast with millions of people trying at once to use the same sofa, or to eat the same plate of paella (Baker 2007). Moreover, these special, communally shared artifacts are in the first instance important constituents of consciousness. Media texts may be construed as the informal textbooks that detail how to be a member of a nation, its culture and sub-cultures. In the case of news media in self-proclaimed democracies, journalism serves a further function as a primer on being a political subject who is supposed to be fluent on issues and participatory.

The stakes are high in properly calibrating media power and accountability. To take one example of that power, on the night/morning of 7/8 November 2000, US Vice President Al Gore was restrained by a body guard from making an electoral concession speech to a live audience in Tennessee (Tapper 2001). As lofty as his position was, Gore assumed he had lost the election before any recounts (later stymied) because he saw it reported as such on television, along with millions of others! Beyond media's power and authority, and following deeply inscribed structural logics, the Edward J. Herman & Noam Chomsky "Propaganda Model" (1988) posits that US news regularly fails to fulfill its remit. In the US, news assumes the dual quality of being at once a commodity (defined as that which is bought and sold) *and* a cultural text; and it performs the double role in a manner that enables tendencies of the former to generally prevail over the latter.

As James Curran observes, it is part of the civic religion in the US that media, including its journalism sector, "be organized as a market, not a state system" and that it be "staffed by professionals seeking to be accurate, impartial and informative" (2011: 1). The model presents attractions, Curran notes, particularly when news media across the globe is often characterized by dysfunction: legalistic harassment (he cites Saudi Arabia), semi-official vigilantism (Russia), top down demands for party line enforcement (Albania, Morocco), tabloid carnival excess (United Kingdom), and near complete sector domination by one politically-active mogul (Italy).

On closer examination, however, the US system has proven "only semi-independent of government" (Curran 2011: 1), despite media's almost ritualistic poses and rhetoric on the issue. Moreover, if news is to be measured by its impact on audiences, the national school room of US news is surely on collective probation. It is important to notice that the Propaganda Model discussed in this volume is not a media "effects" model; rather, it emphasizes the encoding of news messages. Nevertheless, figures on US audiences are bracing and give force to the Herman & Chomsky model's assumptions of news media dysfunction. An example: Justin Lewis, Michael Morgan & Andy Ruddock (2007) report that, in 1992, 44 percent of polled Americans believed that the Bush administration had imposed sanctions on China after the Tiananmen Square massacre. And 43 percent opined (correctly) that the administration had not. The figures may indicate that, remarkably, *zero* percent of US respondents knew the correct "yes" or "no" answer since the same effectively 50-50 split would be achieved by 100 percent guessing or by flipping a coin. By contrast, if 75 percent had answered correctly, it could indicate half knew (50 percent) and half guessed correctly (25 percent) from the two response options.

More recently, in an international comparison of straightforward questions about news topics (Who is Kofi Annan? Who is Nicholas Sarkozy?) US respondents' command of information lags far behind peers in the United Kingdom, Finland and Denmark (Curran 2011: 53-54). The trend of a startlingly ill-informed nation continues. In 2012, Benjamin Valentino reports that 32 percent of US respondents believed that "Iraq had weapons of mass destruction when the United States invaded in 2003" while another 26 percent answered that they did not know. A plurality (not a majority) of 43 percent appraised the claim as false (Valentino 2012: 26). In the same data set, 26 percent maintain that President Barack H. Obama was born outside the US (thus legally barred from the office) while another 19 percent claim agnosia on the question. Among Republicans, a majority hold that Obama was born somewhere beyond the US' perimeter (Valentino 2012: 26).

As it is not reasonable to assume that people in other countries are simply smarter, the question is begged as to why European citizens are better informed than those of the US—and by wide margins, despite the US' collective ego investment in its news media prowess. As concerns the stakes, Curran comments with notable restraint that, "A democracy needs to be properly briefed to be effectively self-governing" (2011: 2). C. Edwin Baker corroborates when he concludes that "a country is democratic only to the extent that the media, as well as elections, are structurally egalitarian and politically salient" (2007: 7).

Writing in the milieu of the 1980s "New Cold War," Herman & Chomsky open their investigation by noting that media operated by state bureaucracies beholden to a party (*i.e.*, the contemporaneous Soviet Union and its satellites) predictably produce the news narratives demanded by power center institutions. By contrast, Herman & Chomsky posit that far more mystification surrounds US news media. This mystification settles notably around the degree to which US news media are subject to (often subtly expressed) State and corporate authority. To further complicate the picture, US media demonstrably "compete, periodically attack and expose corporate and governmental malfeasance, and aggressively portray themselves as spokesmen [sic] for free speech and the general community interest" (1988: 1). Even if it is not the full story of news, upsurges of journalism to civics lesson specifications do happen and breathe life into mythologies about a US press with oppositionalism written into its DNA. But what, more specifically, is the Herman & Chomsky Propaganda Model?

The Model

In crafting their structural model of news media, Herman & Chomsky propose five filters. These five constraints condition news workers at each step in the production process. In particular, the model posits conditioning of news through

1. "the size, ownership, and profit orientation of the mass media" (1988: 2);

2. advertising as the unofficial "license to do business" (1988: 8);

3. sourcing patterns within the regime of professional routines;

4. "flak" with its ideological "enforcers" (1988: 16); and

5. anti-communist ideology as a control mechanism.

They claim that, with these powerful constraints in place, the narratives produced in the news factory reinforce elite-driven "premises of discourse and interpretation, and the definition of what is newsworthy in the first place." As with other forms of enculturation, the filters impact news workers "in such a fundamental way, that alternative bases of news choices are barely imaginable" (1988: 1). Herman & Chomsky also discuss socio-political "dichotomization" that informs news narratives; in turn, their discussion of dichotomization will be important for modifications that I will later introduce to their model.

As concerns the first filter of ownership, Herman & Chomsky orient to "effective concentration in news manufacture" (1988: 4). While there are an array of media firms and channels, down to local papers and radio, Herman & Chomsky observe that the industry's top tier of conglomerates has extensive reach with the power that this engenders. The conglomerates' dominance is evident in commanding vast audiences and articulating the news narratives that gain resilient legs from day-to-day repetition. Herman & Chomsky posit that, to an increasing degree during the 1980s, media firms were becoming more "fully integrated into the market" via greater exposure to financial logics (1988: 5). Loosening of regulatory frameworks were countered with the "sharp constraint" of business-oriented management regimes (1988: 8). Industrial giants with core production far removed from media—Herman & Chomsky cite military-industrial behemoth General Electric—were also emerging as significant media proprietors.

The second filter that Herman & Chomsky propose is that of advertising when it constitutes media's main revenue stream. In the US, ad accounts are far more substantial than revenue from sales of newspapers or magazines while still-prevalent network television has been underwritten in full by commercial sponsors since its inception. One implication of this filter is that ad revenue may trump audience size as a business consideration. This, in turn, empowers content geared toward audiences with higher disposable incomes. Classism is programmed into such a system, as would be obvious in the case of voting weighted by income. Herman & Chomsky observe that monetary "votes" from advertisers have proven more effective than harassment from governments in shuttering radical and class-conscious news media by drying up ad revenue. As a further concomitant of courting advertisers, media firms tend to "avoid programs with serious complexities that interfere with the 'buying mood'" that is preferred by sponsors (1988: 11). Tamer "legitimate controversies" or personalization of political issues readily infuses the news hour, on these logics.

Sourcing presents the third filter that shapes the contours of news media. Herman & Chomsky observe that "bureaucratic affinity" is one factor that leads news organizations to gravitate toward other bureaucracies—State and private—as the leading founts of front-page issues, backstory, and quotes. Moreover, to station reporters at these bureaucratic centers from which news regularly emanates is cost effective deployment of resources. Hence, the White House correspondent reports for duty, to be assisted by the White House press office. Information obtained from bureaucracies is further advantaged for being taken as "presumptively accurate" and without demands for explanation of why it is newsworthy (1988: 14). As valued conduits of information, these regular sources command a degree of deference that they may not merit: "It is very difficult to call authorities on whom one depends for daily news liars," as Herman & Chomsky observe (1988: 14). The rise of think tanks after the 1970s reinforces the sourcing filter. Think tanks convene and may co-opt ostensible experts to act as "purveyors of the preferred view" when funded by interested foundations (1988:16).

Herman & Chomsky employ the term *flak* in introducing their fourth filter. Flak may be more or less massive and coordinated, while it is enacted through "negative responses to a media statement or program." In turn, negativism gains expression via "letters, telegrams, phone calls, petitions, lawsuits, speeches and bills before Congress, and other modes of complaint, threat, and punitive action" (1988: 16). "Serious flak" may take the form of lawsuits and other attacks that can lead to the ouster of an individual from his/her position or disable an entire organization. It bears notice that, beyond the instruments of flak available at the model's formulation in the 1980s, there are now many more arrows in the quiver (for example, tweets, abusive web pages, denial of service attacks). While the remit of flak extends far beyond the news, Herman & Chomsky argue that its purpose is often pre-emptive as it "conditions the media to expect trouble (and cost increases) for violating right-wing standards of bias." Interestingly, "although the flak machines steadily attack the mass media, the media treat them well" via "respectful attention" (1988: 18).

Anti-communism presents Herman & Chomsky's fifth filter. Through 1988, widely circulating anti-communist ideology enabled a US foreign policy that empowered abusive right-wing dictatorships (notably in Latin America) as presumptive bulwarks against "the commies" (typically, pauperized urbanites and landless peasants). Anti-communism also served the significant purpose of placing the non-communist domestic left (unions, centrist liberals, intellectuals) on perpetually defensive footing. Following the rollback of command socialism at the conclusion of the 1980s, this filter is no longer of much moment, although some dead-enders such as the flak organization Accuracy in Media perseverate with anti-communist paroxysms (Goss 2009). Thus, I have retrofitted the more general mechanism of dichotomization that Herman & Chomsky extensively discuss over their original anti-communist filter. The retrofitting is appropriate since anti-communism had the features of a classic dichotomized discourse on the Other. Furthermore, the discursive mobilization of "Us versus Them" dichotomies is undiminished in the present, with new brigades of threatening (internal and external) Others conscripted for duty.

The remit of this volume is to assess the model, first published in 1988, in its fit to current times. To what extent have world historic developments, from the demise of command socialism to the advent of the internet, complicated or impacted the model? The short answer is that, despite the hurly burly of superstructural change, events of the past quarter century have largely if unevenly reinforced the functioning of the model's filters. In elaborating this thesis, I am attentive to noteworthy developments that circumvent the reach of the filtering mechanisms—and am also alert to the ways in which new media have been recruited into reinforcing (and not, as widely assumed, monolithically upsetting) business as usual.

Evaluating the Model

Joan Pedro (2011) exhaustively reviews scholarship on the Propaganda Model, with some stress on unsympathetic reception toward it. Indeed, the literature that has developed on the model ranges from hagiographic to hysterical. At the former pole, Jeffrey Klaehn's account devoutly reads one long Chomsky quotation after another into the record, in lieu of further probing (2005). As for the latter pole, Michael Schudson strenuously attacks the Herman & Chomsky model as "misleading and mischievous" (2011: 31). Nonetheless, Schudson subsequently befogs the issue with unmarked vacillations between red herring distractions and *sotto voce* acceptance of some Herman & Chomsky premises. In the light of critical division, it stands to reason that researchers' stances toward the Propaganda Model parallel their omnibus judgment on the performance of US news media. In this vein, Schudson posits that the news about US news is good indeed. Defying restraint, he rhapsodizes that "the genius of American journalism is the symbiosis of professionalism and commercial organization." By contrast, W. Lance Bennett's (2001) more critical and trenchant appraisal of US news media is favorable to the Propaganda Model even as he relies on other lines of analysis.

The Question of Determinism.

The Herman & Chomsky model necessarily posits that the work of a journalist is circumscribed by an extrapersonal industrial organization embedded within a corporate-dominated order. These conditions furnish the "always already" at hand practices of culture that continuously massage news narratives into form. Notice that assumptions (about, for example, a system's form and function) are unavoidable in constructing or in taking up a model. Indeed, the very reason that investigators construct models is to tease out and explain the massive and predominant tendencies of a complex system's behavior.

It is debatable, however, whether Herman & Chomsky have fashioned a strongly determinist model that only permits limited autonomy between its constituent parts. Shades of determinism are evident in Herman & Chomsky's muted interest in cases of "deviant journalism" that arise even within the mainstream. As concerns the question of determinism, it is important to bear in mind that social systems do not exhibit the same mechanics as many of those in the natural sciences. In this view, it is not merely glib for Berelson to construe the impact of media in notably qualified terms: "Some kinds of communication on some kinds of issues, brought to the attention of some kinds of people under some kinds of conditions, have some kinds of effects" (quoted in Diamond & Bates 1991: 347).

Given the degrees of autonomy that arise in social and institutional behavior, I am attentive to news discourses that go against the grain. Such moments in the news are discursive "hiccups", exceptions that prove the rule of discourse that follows the legitimate controversies. Against-the-grain moments may, however, auger a rupture when the definition of what is, and is not, marginal in news undergoes a recalibration. Against-the-grain news is understood to occur at the micro-level of particular stories as will be noted throughout Part II of the volume. At the macro-level of a well-known and mainstream media organ, in Chapter 5, I extensively employ *The* (London) *Guardian* as a case study of a medium that resists some of the filters, even if it too is part of a flawed news milieu.

While interested in the meaning of exceptional cases, I ground my analysis in the Propaganda Model for the straightforward reason that I posit news as having a significant degree of constraint driven by the conditions of its production. Following previous investigations of US news media behavior (2001, 2002, 2003, 2009), I conclude that there is a problematic reality in its performance that demands to be addressed. In the light of structural impacts of, for example, ads and sourcing, the news is not the outcome of a shapeless and anarchic milieu. Rather, as amplified through the model, news is the industrial product of "structuring structures" (Benson & Neveu 2005: 3), exerted day after day and down into the details of journalistic practice. In positing the news product as relatively but not absolutely autonomous from the materially grounded prerogatives that surround and permeate it, one rejects a mechanistic form of structuralism more suited to account for the behavior of photocopiers than that of human endeavors. In this view, it is nevertheless nonsensical to assume that structural (economic, organizational) demands do not impact regularly and significantly on news narratives; hence, the appeal to a model to capture, order, and explain these structures that condition news performance.

Departures

This volume presents distinctions from the seminal Herman & Chomsky model. As noted above, a dichotomization filter has been retrofitted over anti-communism since new Others are playing the part of Them (distinct from Us) in contemporary news discourse. Furthermore, I extensively address new media throughout the volume as well as in its own dedicated chapter. Finally, in a move that Herman & Chomsky do not venture, I apply one filter to news discourse beyond the US at chapter length. In taking the model offshore to the UK, I do not preemptively dismiss objections on the grounds of over-generalization as merely the narcissism of small differences. The UK *is* substantially different from its former colony the US![1] However, part of the point of Chapter 5 is that differences are consequential when *The Guardian*, partially buffered from market relations, fashions news discourses that are less beholden to the moralistic and populist inflections of Us/Them dichotomization. I am dismissive, however, of manipulative and defensive expressions of nationalism that posits one's own nation as immune to Us/Them binaries. The London-based *Daily Telegraph*'s mobilization of Us/Them discourses furnishes clinching evidence on this score.

Concurrent with my efforts to update Herman & Chomsky, the Propaganda Model may be assuming greater salience internationally since the US' media template can and is being exported beyond its own perimeter. Neoliberal drives toward privatization of media have been a significant export since the 1980s (Garnham 1990), even if other Western states do not approach the level of commercialism evident in the US where it is effectively "game over." The fact that the model of US news media demonstrably "works" in shepherding elite socio-economic interests into the mainstream without an overtly heavy hand is, moreover, surely not lost on other nations' potentates.

Part I of this volume is entitled "Contexts" and it will unpack the "infrastructure" of the model's first three filters: patterns of ownership, the impact of advertising within a broader commercial milieu, and sourcing as embedded within contemporary journalism's professional routines. Whereas Herman & Chomsky cover the first three filters more briefly before they move to their extended case studies, I dwell at length on the ways that the trio of ownership, ads, and sourcing shape the practices of journalism. While evidence will be mainly drawn from the US news industry, international (mainly UK) and recent examples will be marshaled. I argue that the filters are at least as powerful as they were when the model was first proposed. For instance, ownership can be construed as not only more concentrated but more sensitive to market signals while the internet has, if anything, raised the tempo where news speed-up is concerned.

The second half of the volume, entitled "Texts," will address a series of case studies of news media discourses that are products of the first three filters' infrastructure. The discourses are further assumed to be informed by dichotomization, expressed through the Us/Them binary, and by the production of flak. In particular, Chapter 4 outlines a theory of Us/Them that expands out the Herman & Chomsky discussion. It is a case study of Us/Them dichotomization; the chapter discusses the recent nadir of journalistic performance in *The New York Times*' discourse on the US and United Nations during the run-up to the 2003 invasion of Iraq. Chapter 5 also takes up dichotomization and revolves around a paired case study of two British broadsheets, *The Guardian* and *The Daily Telegraph*. In covering the 2011 riots that convulsed England, one paper readily slid into Us/Them dichotomies while the other assayed to channel far more of the texture and backstory of events.

Chapter 6 examines flak via a dedicated producer of it, the Virginia-based Media Research Center (MRC). A reading of MRC's recent "Special Reports" demonstrates that the organization manufactures junk research. However, MRC's remit is not for investigative rigor, but to insinuate its contrived findings into other media channels for the purpose of stimulating flak episodes. Finally, Chapter 7 addresses the still unfolding impact of new media. While I entertain heavy criticisms of the new platforms, an examination of the self-reflexive domain of news media criticism yields an interesting result. Whereas incumbent media's foremost critic of journalism, Howard Kurtz, steadfastly defends received journalistic wisdoms and practices, Glenn Greenwald assertively cross-examines news media performance from a new media platform.

Finally, in minting all of these claims, I am cognizant of the journalistic heroes who define the profession at its best—both the relatively high-profile figures (discussed by Pilger 2005) and the less hyped ones. Indeed, the miracle of the newspaper is an essential part of the quotidian in which I eagerly participate. Far from condemning journalism, I hope to see it more closely approximate its mythologies about itself. The clarity and quality of our consciousness and trajectory into the future depend upon it.

Notes

[1] Notice that the otherwise valuable Stuart Allan edited volume *Journalism: Critical Issues* (2005) ranges across the English speaking world—mainly United States and United Kingdom—with unmarked continuities and discontinuities between national journalistic cultures.

PART I
Contexts

CHAPTER 1
Owning the News Discourse

On Saturday 2 July, Rupert Murdoch's daughter Elisabeth and her millionaire PR husband Matthew Freud hosted a party at their 22 bedroom mansion in Cotswolds. Michael Gove the education secretary was there. So was [Prime Minister] David Cameron's consigliore Steve Hilton, and the culture minister Ed Vaizey. The Labour figures in attendance included Peter Mandelson, the ex-work and pensions secretary James Purnell, the shadow foreign secretary Douglas Alexander—and his shadow cabinet colleague Tessa Jowell. (...) They were joined by [former Foreign Minister] David Milliband. (...) The BBC's director general Mark Thompson turned up [along with a slew of media notables] (...). Also among the guests was [News Corporation director] James Murdoch who spent much of the night talking intently to Rebekah Brooks [Chief Executive of News Corporation subsidiary News International].

—John Harris (2011: G2:5)

In the celebrated 1941 film, *Citizen Kane*, Orson Welles and his filmmaking colleagues conjure a situation in which an unaccountable media mogul uses his image empire to joust, arrogantly but clumsily, against reality. The invention of Charles Foster Kane was not entirely fictitious as the film included references to contemporaneous media mogul William Randolph Hearst. The thrust of the film's liberal politics are, nevertheless, softened by collapsing into a portrait of a merely psychologically arrested and privately inept figure whose political machinations collapse.

Real life has proven far grittier than *Kane* in the case of Keith Rupert Murdoch, the Chairman and CEO of News Corporation.

Murdoch has constructed a media empire across more than a half century, beginning from modest media holdings in Australia. News Corporation has been characterized, perhaps hyperbolically, as the only "truly" global media firm (a view endorsed by Flew 2007: 82) among a number of gigantic North American, European, and Japanese firms that operate extensively beyond the nations in which their home offices are domiciled. *Columbia Journalism Review*'s "Who Owns What" feature on its web page surveys the extraordinary depth and geographical breadth of News Corporation's reach (*Columbia Journalism Review* 2011). Alongside behemoth size, Murdoch's management style has been forcefully criticized for keeping authority literally "in the family" through its bloc of voting shares, making "obscenely nepotistic" decisions, and stiffing investors via the company's underperforming returns (Smith 2011: 24).

Beyond media businesses, News Corporation is a player in politics. Lance Price, a former media advisor to Britain's Prime Minister (1997-2007) Tony Blair, characterizes Murdoch's influence during the "liberal" Blair era as follows:

> When I worked at Downing Street, he seemed like the 24th member of the cabinet. His voice was rarely heard (...) but his presence was always felt. No big decision could ever be made inside No 10 without taking account of the likely reaction of three men—[Chancellor of the Exchequer, later Prime Minister] Gordon Brown, [Deputy Prime Minister] John Prescott and Rupert Murdoch. On all the really big decisions, anybody else could safely be ignored. (2006: 1)

Alongside its penchant for some tawdry content, News Corporation clearly takes consistent stands on some "really big decisions" across its vast media holdings. For example, in the run-up to the invasion of Iraq in 2003, global opposition was intense, from elites to the streets. No similar controversy animated News Corporation publications. In a stunning display of discipline, 247 out of 247 Murdoch papers around the world editorially lined up behind the dubious cause of the invasion (Harvey 2005: 35). Another characteristic of Murdoch's career has been indulgent service from governments regardless of regulatory statutes on the books. In this vein, News Corporation exceeded even liberalized post-1996 US limits on national television penetration by the turn of the millennium (Wexler 2005: 14).

In practice, however, what difference does media ownership make? With gale force wind at challengers Tony Blair and Barack Obama's backs in their respective 1997 and 2008 electoral campaigns, the tired and discredited incumbent ruling parties effectively faced no chance regardless of media performance. However, when electoral matters are up for grabs, Murdoch's News Corporation has not been timid in putting its thumb on the scale. During the election night/morning of 7/8 November 2000 in the US, Fox News projected George W. Bush as the victor based on partial returns in an astonishingly close contest. As a result, the other networks unwisely rushed to follow suit, even as Florida's mandated automatic machine recounts had not even begun. Thus, millions of citizens went to bed late that night believing the outcome settled, while Gore had to be restrained from making a concession speech. In the all-important contest of impressions and procedures that followed, the narrative that readily congealed posited Gore as vaingloriously trying to annul a *fait accompli* (Goss 2003). And who made the momentous decision for Murdoch's Fox? It was John Ellis, chief of Fox's decision desk—and first cousin of candidate Bush. Moreover, the cousins conversed intensively on the phone during the election night (Wittstock 2000). Fox's decision to install Ellis as chief of the decision desk was breathtaking in its disregard for elementary ethics surrounding conflict of interest, a charade of the sort that one would expect to find in a nation that lacked mature democratic institutions such as independent media.

The opening vignette of this chapter describes a more recent scene from the summer of 2011. The *mise-en-scene* in the rarified Cotswolds testifies to the reach of Murdoch's News Corporation into the exclusive club of the governing and cultural elite. Coziness prevailed up until the moment that News Corporation's phone hacking scandals broke open on the following Monday, 4 July. As the scandals began to gain legs as a news story, Vincent Cable commented on the mood among his colleagues in British Parliament in terms that are unusual in a Western republic. It was, quipped Cable, akin to "the end of a dictatorship, when everybody suddenly discovers they were against the dictator" (quoted in Murphy 2011).

Subsequent fallout from the phone hacking scandals has included more than two dozen arrests of people implicated in the activity (Halliday & Dodd 2012), the abrupt shuttering of Murdoch's hugely successful *News of the World* tabloid in July 2011, and large pay-outs to 37 victims of phone hacking in January 2012 after the judge found "'a previously conceived plan to conceal evidence'" (presiding Mr. Justice Vos, quoted in Sabbagh & Hill 2012: 1). In a striking concession, News Corporation agreed to pay fines "'as if' the allegations that they lied, obstructed police, and destroyed evidence were true" (Sabbagh & Hill 2012: 1). If power corrupts to a degree in line with the extent of that power, News Corporation's behavior betrays the enormous entitlement that comes with long getting its way in the mediascape.

Decades after *Kane*'s provocative but attenuated critique of media power and before News Corporation's current straits, Ben H. Bagdikian's *Media Monopoly* was first published in 1983. Bagdikian estimated that, at the time, there were 25,000 media firms in the US (newspapers, magazines, radio stations, book publishers, and major movie studios). While this seems like a media oasis of dispersed discursive authority, Bagdikian posited it as more like a mirage since there were not "25,000 different owners" (1992: xxix). Indeed, Bagdikian claimed that fewer than 50 firms were capturing more than half the revenue from all media combined (1992: xxviii, 21). By the fourth edition of the book in 1992, following the Reagan-Bush wave of neoliberal deregulation, the set of media firms that commanded more than 50 percent of media revenue had winnowed down further to 20 (1992: ix). The specific numbers have been contested by scholars who are sympathetic (Baker 2007: 54-55; Noam 2009: 20) and antagonistic (Compaine 2004: 2-3) toward Bagdikian. However, Bagdikian's work at least propelled ownership into the discussion of mass media. The issue was subsequently picked up in Herman & Chomsky's formulation of their model.

Beyond numbers that testify to bigness, Bagdikian's *Media Monopoly* expresses concern that the media has been folded into larger conglomerates at the heart of the "military-industrial complex" about which former five-star General Dwight D. Eisenhower warned in his final address as US president (Eisenhower 1961). Bagdikian dwells on General Electric (GE), a manufacturing colossus, large-scale military contractor and, then as now, proprietor of the National Broadcasting Company (NBC). GE is wedded to government through military contracting, campaign contributions and lobbying (Bagdikian 1992: 11; also see Greider 1992: 331-355). The company is, in turn, interlocked with other industrial giants via over-lapping directorships (Bagdikian 1992: 24). Within this political economy, conglomerates such as News Corporation and GE assay to influence the political climate by "own(ing) most of the news media" (1992: 26) and shape their economic destinies through the risk-reduction strategy of commercial synergy across their corporate divisions (1992: 243). It (almost) goes without saying that the interests of these companies strongly favor deregulated economic neoliberalism that encourages their hypertrophy.

The question is thusly begged of what, if anything, has changed since Herman & Chomsky fashioned the Propaganda Model. Does the first filter of the model still hold? Or have the dispersed forces of globalization and the "communications revolution" with its new media substrates rendered the proposed ownership filter as (totally? partly?) obsolete? Was comeuppance for News Corporation foreordained by a now passing age of corporate dinosaurs?

In beginning to sketch answers, one may posit that ownership is surely more remote in its impact on journalism than news workers' routines that are proximal to filing a news report. However, as this chapter will attempt to demonstrate, ownership patterns shape the ecological niche in which journalists work in the first instance. While Herman & Chomsky may have been precocious in visioning media ownership as highly concentrated, this chapter examines the current state of empirical evidence. Moreover, I flesh out the theoretical arguments that affirm dispersed ownership as a pillar of liberal self-governance congruent with Enlightenment ideals. I also discuss other developments since the Propaganda Model's original 1988 publication. In particular, institutional investors and a stricter bottom-line business orientation are now more prevalent, while moguldom is in relative decline. Before addressing these issues in detail, I describe and critique the neoliberal economic paradigm that initiated its ascendency in the 1970s and has furnished the framework in which news and culture are produced.

The Big Picture: Neoliberalism

Since the 1970s, economic paradigms have listed toward "Neoliberalism [that] has, in short, become hegemonic" (Harvey 2005: 3). Neoliberal doctrines have been boosted by concerted propagandizing on their behalf with media-savvy think tanks as a leading fount of carefully tailored facts and arguments (Goss 2000a; Soley 1995). At the same time, more than arcane formulas are at stake as economics conditions the allocation of wealth and opportunity. Furthermore, as Margaret Thatcher aphorized about this particular academic discipline's more profound registers: "'Economics are the method (...) but the object is to change the soul'" (quoted in Harvey 2005: 23). In this view, different economic theories produce their own distinct kind of subject when unleashed on whole societies. As concerns the neoliberal subject, he or she is saturated in commodification and individualism within the social Darwinist milieu that unbridled markets promote (Goss 2000b).

The policy package of neoliberalism includes rollback of union power, a stripped-down welfare state, and tight money monetarism to police inflation. Moreover, ambitious privatization programs open even notably efficient public infrastructure—water, phone systems, trains, schools—to market levels of pricing and service (often observed to be high and low, respectively). Further features of the package include reduced taxation on wealth, aggrandizement of investment opportunity with far fewer brakes on capital mobility, and light touch regulation over complicated financial products (Harvey 2005: 33). Through it all, the State is posited as an irreducible antagonist to the market. Thus, under the neoliberal prescription, the State's social welfare functions are rolled back and the population is far less buffered from the often severe vicissitudes of market discipline.

A measure of dynamism has followed neoliberalism's implementation since new industries gain ascendency, speculative wealth metastasizes, and even the pace of new products and fashions increase in tempo (Harvey 1989). It is a veritable thrills and chills ambiance of the vaunted and convulsive "creative destruction." With the emphasis on restricted growth in the money supply (monetarism), the neoliberal prescription has also been generally effective in arresting inflation while keeping real rates of interest (and levels of debt) at high levels. When one looks at the whole scorecard, however, neoliberalism has exhibited severe defects. Neoliberalism's record for economic growth (very low), income distribution (highly skewed), debt (exploding for individuals, corporations and governments) has been strikingly poor as compared with the pre-neoliberal era before 1980. Moreover, the tendency toward severe crises of a sort that were exceptional from the 1940s into the 1970s has greatly intensified (Harvey 1989: 141-172). Although government activity is openly despised by neoliberals when it generates efficient and widely used public services, it is nonetheless ratcheted up (if at times inadvertently) under neoliberalism. Stepped-up government activity occurs via bailouts to rescue "too big to fail" firms, industries or whole national economies during episodes of meltdown.

Despite its failures in practice, neoliberal exponents' indignant utopianism is undaunted as they exalt the doctrines' ostensible promotion of "freedom" and "prosperity" (Friedman & Friedman 1980). Emollient words notwithstanding, the central strategy of neoliberalism goes "back to the future." Under the banner of neoliberalism, a sustained frontal attack has been waged on the post-World War II compromise between capital and labor (Keynesianism, also known as "Fordism" [Harvey 1989] or "embedded liberalism" [Harvey 2005]). The post-WWII Fordist compromise implicated "increases in purchasing power, policies in favor of full employment, and the establishment of the so-called welfare state" with substantial insurance against illiteracy, unemployment, sickness, and old age (Duménil & Levy 2011: 16). In Duménil & Levy's terms, the Fordist managerial class that mediates between capitalist and popular (working) classes was more aligned with the interests of the latter during the era of compromise. At the same time, Fordist governance continued the State's longstanding ministering to markets via the provision of extensive infrastructure, research and development, and the cultivation of "human capital" via education. While Fordism's package of programs generated broadly spread benefits and an appreciable measure of social solidarity, it also presented concomitant curbs on the reach of the very wealthy and straight-jacketed casino-like financial activity.

In their monumental work that sifts through the structure of neoliberalism's performance, Duménil & Levy conclude that it "expresses the strategy of the capitalist classes in alliance with upper management, specifically financial managers, intending to strengthen their hegemony and expand it globally" (2011: 1). Moreover, the neoliberal regime crystallizes "a social order aimed at the generation of income for the upper income brackets, not investment in production nor, even less, social progress" (2011: 22). By the end of the first decade of neoliberalism's implementation in the US during the Reagan era, empowerment of the wealthiest class fractions was evident (Harrison & Bluestone 1988: 21-52, 109-138) while an entire deregulated industry (Savings and Loans) was strangled by the "invisible hand" and in need of resuscitation by government bailout (Pizzo, Fricker & Muolo 1989). During the 1990s, the neoliberal regime appeared to achieve convergence between elite and popular interests. Into the present, however, neoliberalism has unbalanced the global economy and generated the perfect storm of volatile conditions that prefaced the global crisis that began in 2008.

As noted, the State is the villain in the neoliberal grand narrative. Friedman & Freidman insist that, "Every government measure bears, as it were, a smokestack on its back" via damaging externalities (Friedman & Friedman 1980: 32).[1] However, far from rejecting the State as in Friedman & Friedman's intellectually dishonest narrative, neoliberalism is a prescription to retool the State's activities. While backing off from social investment and safety nets, the neoliberal State intervenes in the interests of the wealthiest class fractions by fashioning tax breaks as well as direct and indirect corporate subsidy. The coefficient of State activity shifts from redistribution through public infrastructure (schools, public transit, public health) to coercion that polices and carceralizes the resultant social disaster. Doctrines that have been dogmatically couched in the rhetoric of "freedom" depend on the iron fist of the State's most coercive activities to be implemented outside the perimeter of the gated community.

Implications follow for media industries. The most obvious is that ownership concentration is enabled via mergers and acquisitions when regulation is rolled back in favor of neoliberalism's formula for skewed wealth distribution. With respect to media content, it is also more likely to channel the elitism and consumerist celebration of the neoliberal zeitgeist (Goss 2000b; Hackett 2005).

Media Industries: A Primer

Despite its enhanced muscle, media is not (nor is it likely to become) the "core" of economic activity and wealth production: "The size of even the largest media entities pales beside the legally permitted size of oil companies and banks, much less Wal-Mart" (Baker 2007: 14). While information firms are among those with the highest market value in global terms—for example, Apple rests at number one with a value that approached $US600 billion by the end of 2011—media content producers barely nose into the Top 100. According to *Financial Times* (2012)'s "Global 500 2012", the only media production companies that reached the top 100 firms in market value were Comcast (Number 71) and Disney (Number 74) in a list that is laden with banks and heavy industries. Nevertheless, as noted above, media industries "punch above their weight" in cultural terms and they also exhibit some of the same patterns of the larger economy in which they are immersed.

In more specific terms, media firms have clear incentive to conglomerate given that so many texts fail to find an audience. As Flew observes (2007: 12), demand for texts is highly elastic and may effectively reach zero in some instances. In this vein, consider the difference in the levels of demand for Dan Brown's cultural commodities when he was a singer-song writer at the start of the 1990s (almost no demand) versus the frenzied public appetite for his books in the decade that followed (Rogak 2005). In light of notably uncertain demand, aggregation into larger industrial combinations hedges commercial bets so that the many market failures spawned by a media firm can be effectively cross-subsidized by the relatively few successes.

The increasing size of information sector firms (Noam 2009) has been propelled by merger mania that has intensified since the

1980s. For instance, Germany-based Bertelsmann was involved in 386 acquisitions as buyer or seller between 1992 and 2002—a dizzying pace of more than one deal every ten days across the decade (Hesmondhalgh 2007: 162). The cascade of acquisitions enables synergy that can be construed as placing "'a brand at the hub'" of a commercial wheel "'with each of the spokes a means of exploiting it'" (*The Economist*, quoted in Wasko 2001: 71). As Wasko details at length, Disney's synergies for a typical film include: themed apparel, gifts, home furnishings, toys, sporting goods, publications, public events (parades, mall tours), synergistic references in its other media such as ABC network, among other commercial reverberations. In this case, the cascade of synergy was all convened for the largely forgotten 1998 film *Hercules* (Wasko 2001: 70-83).

Within this media ecology, Hesmondhalgh identifies seven firms as comprising an echelon of their own by 2005. The Seven Titans are, in descending order of revenue in billions of US dollars: Time Warner (43.7), Walt Disney (31.9), Viacom (27.0), News Corporation (23.9), Bertelsmann (21.6), Sony (16.0), and NBC Universal (14.7) (2007: 163). Interestingly, each firm is based in the US, Western Europe or Japan, the tripartite zone of economic wealth across the long sweep of recent generations. Hesmondhalgh observes that beyond the Seven Titans, there is a second tier of several dozen important media firms that may burst into the top tier as the industry churns. That was the story of Viacom. Originally spun off by its parent company CBS in 1970, Viacom's subsequent rise enabled a merger with its erstwhile corporate parent by 1999.

Beyond the Titans and the second tier, there are thousands of small firms that sub-contract and collaborate with the larger firms. Hesmondhalgh claims that, "Such companies may account for small levels of market share, but they are important in terms of the numbers of people they employ and their potential to foster...innovation" (2007: 175). Consider the film industry. The firms in the major studio oligopoly (Warner Brothers, Twentieth Century Fox, *et al.*) often collaborate with small firms in co-productions. This explains the legion of title cards that typically appear before the start of a film that recognize the firms with a stake in the production. More generally, small and large firms can be ensconced in "interdependent webs" that are also known as "alliance capitalism" or "co-opetition" for their admixture of competition and co-operation (2007: 176-77). The upside for firms to engage in "co-opetition" include shared (reduced) risk and seats on ostensible rival's boards that present a buffer against market turbulence. Smaller firms may also be formally incorporated into a conglomerate by being vacuumed up as acquisitions.

Strictly speaking, the situation is not as monolithic as implied in the phrase "media monopoly." Nevertheless, the cultural concomitant of all this activity in the corporate tree-tops has been felt down below. Hesmondhalgh perceives "unprecedented commercialization of our everyday cultural lives over the past 20 years" (2007: 278). Intensified commercialism is a result of, and further intensified by, media regulatory policies that I discuss next.

Further Implications of Neoliberalism for Media

Neoliberalism's march across the globe has concomitants in policies toward media industries. While long characterized by its predominantly private media system, the US has more recently wielded the carnivalesque "soft power of its global attraction" in favor of a global privatization wave (Curran 2011: 1; also see Garnham 1990). Since the 1980s, the US has also stepped-up the domestic commitment to privatization in media via heightened deregulation and a permissive framework for ownership concentration. The US' Telecommunications Act of 1996 was neoliberal in spirit and enacted a sweeping package of deregulatory policies across radio, television, and telephony. Prior to the act, the limit on radio stations that a single firm could own had been set at 40. In a radical reversal, the act removed any upper limit and merely capped ownership at eight stations in any one city. After the act, merger mania gripped the radio industry and, by the end of the decade, Clear Channel had amassed 1,200 stations (30 times the previous limit). Clear Channel alone captured 42 percent of listeners and 45 percent of all US radio revenue (Wexler 2005: 10). The advent of unchecked concentration in the industry prompted far less local news and far more homogenous formats and playlists. Market power also enabled radio giants to practice political strong-arming by privileging syndicated right-wing talk radio and to black-list artists who criticized George W. Bush's administration (Jolly 2007; Wexler 2005).

Proposed regulatory changes in 2003 went further still in dismantling of remaining regulatory firewalls such as those between local cross-ownership of broadcast and print. In this instance, public and legal pushback was fierce and prevented implementation of the package (McChesney 2004). Although a next wave of deregulation was stymied, neoliberalism continues to steer the State, its ostensible foe, with pro-wealth and conglomeration policies welded into place. Michael Copps, a commissioner on the Federal Communication Commission during its most fervently neoliberal phase under chair Michael Powell (2001-05), subscribes to the previously assumed and now contrarian view that the broadcast spectrum belongs "to the people." As it is worth "worth hundreds of billions of dollars," Copps argues that the broadcast spectrum should be managed for public interest with government oversight on the private companies entrusted with broadcast functions (2011: 292). Despite these stakes, public interest reviews on stations' performance during license renewal for a slice of the broadcast spectrum "now are less than cursory" (2011: 292).

"Pretty Lucky"

Benjamin M. Compaine, "an ardent neoliberal" (Hesmondalgh 2007: 188), is "possibly the most scholarly critic of the view that existing concentration in the mass media is...objectionable" (Baker 2007: 54). Thus, Compaine's corpus presents an appropriate site from which to examine the implications of neoliberalism for media economics. Most fundamentally, Compaine claims that media plenitude is the "empirical reality" (2005: 46). He enthuses that, "We should consider ourselves to be pretty lucky" with media as it is in the US where privatization and substantial deregulation are the rule (2004: 8).

Compaine flatly characterizes the "basic premise" that US media is concentrated as "false" and points to his calculation of an antitrust index for support (2004: 1). The procedures by which Compaine calculated the index will be taken up later. While seeing no concentration of ownership to speak of, Compaine also vigorously defends Clear Channel's acquisition of 1,200 radio stations following the 1996 deregulation (e.g., 2004: 7).[2] In such moments, neoliberal discourse shifts the ground on which it stands from divining no industrial concentrations, to positing that any private corporation that "wins all the chips" is by definition deserving of them. Furthermore, in defense of the corporate media, Compaine insists that larger firms are likely to produce equally good or better content than that of smaller firms.

Exponents of neoliberalism strain to finesse the deeply inscribed conflict between their avowed theoretical commitments and the empirical realities that those commitments set into motion. In particular, the neoliberal demand for media market deregulation has been "promoted in the name of *competition*" while, predictably and in practice, it has "facilitated *consolidation*" (original emphasis, Boyd-Barrett 2005: 343); and consolidation and its associated corporate hypertrophy necessarily dial down competition. Furthermore, neoliberal complacency about consolidation presents a radical departure from the fundamentals of the founding liberal doctrines. The avatar of liberal capitalism, Adam Smith, wrote scathingly of industrial concentrations in the eighteenth century since they annulled brisk de-centered competition and deviated from the equilibrium of the "natural price" under market auspices (1904: 24-25).

While glossing over the question of dampened competition under a regime of consolidation, Compaine also insists that media conglomerates produce better quality programming as compared with smaller firms. He writes, "The academic research that has been reported does not support the contention that media ownership by chains or conglomerates leads to any consistent pattern of lowered standards, content or performance when compared with media owned by families or small companies" (2005: 10). The claim may even seem intuitive given that larger firms command more resources than smaller firms.

However, the extensive Project for Excellence in Journalism (PEJ) study that Compaine cites for support draws a crystal-clear conclusion about television news: "Our five year data suggests that when it comes to overall quality, smaller is better" (2003: 3). PEJ reiterates that "we found clear distinctions" among ownership categories, as "The smallest companies produced higher quality newscasts" (2003: 7). Surveying its data, the PEJ argue forcefully against the core neoliberal prescription of deregulation and the consolidation that follows in its wake:

> Most importantly, the data raise serious questions about regulatory changes that lead to the concentration of vast numbers of TV stations into the hands of a few very large corporations. The findings strongly suggest that this ownership structure, though it may prove the most profitable model, is likely to lead to further erosion in the content and public interest value of the local TV news Americans receive. (PEJ (2003: 7)

While the PEJ is baldly critical of the concentrated ownership associated with neoliberalism in practice, Compaine devotes three paragraphs to the report plus six bullet pointed highlights of its conclusions—and ignores findings that are unmistakably central, albeit diametrically opposed, to his policy prescription for deregulated media ownership (2005: 10-11). Instead, Compaine rhetorically waves his arms and proclaims that "there is something for everyone" in the report's conclusions (2005: 11).

Measuring Markets: By the Numbers

What degree of concentration (or non-concentration) now characterizes the industries that deal most directly with (political, social) consciousness? The neoliberal Compaine bluntly dismisses any notion that ownership is anything other than widely dispersed in the US mediascape: "The media industry is not, as matter of fact, highly concentrated" (2004: 1). In one reply that is grounded in an exhaustive parsing of data, Eli M. Noam (2009) collates the "Information Sector" into an array of distinct categories ("Mass Media," "Information Technology," Telecommunications," and "Internet"). In turn, Noam identifies and analyzes over 100 sub-industries under the umbrella of the information sector. The straightforward formula that informs the number crunching is derived from the US Department of Justice's Herfindahl-Hirschman Index (HHI). Noam cautions that the HHI is "not a binding rule but a guideline" in antitrust action and should "be supplemented by additional factors and be used with discretion" (2009: 412).

The HHI is calculated as a sum of the squares of market shares within a designated economic sector. The index assumes this form—and not simple addition of market shares—on the assumption that each additional percent of market share that is gained may "generate more than a linear increase in market power" (Baker 2007: 58). Given the HHI's construction, the highest concentration score that can be obtained is 10,000 which corresponds to one firm that gains 100 percent of market share (i.e., 100 squared equals 10,000). While one firm capturing the entire market is clearly extraordinary, how are other scores interpreted? The index indicates a low degree of concentration for an HHI score of less than 1,000. Moderate concentration is defined as an HHI score of between 1,000 and 1,800 while high concentration is construed as a score that exceeds 1,800.

Thus, if a market is constituted by eleven firms that each capture 9 percent of the market, the resultant HHI score of 891 is less than 1,000. The calculation is as follows:

$$(9^2) + (9^2) + (9^2) + (9^2) + (9^2) + (9^2) + (9^2) + (9^2) + (9^2) =$$

$$81 + 81 + 81 + 81 + 81 + 81 + 81 + 81 + 81 + 81 + 81 = 891$$

As the HHI is under 1,000, this hypothetical market is effectively competitive by the numbers. By contrast, a hypothetical market that features five firms that each capture 16 percent of the revenue, and two firms that each garner 10 percent, generates a HHI of 1,480. In particular:

$$(16^2) + (16^2) + (16^2) + (16^2) + (16^2) + (10^2) + (10^2) =$$

$$256 + 256 + 256 + 256 + 256 + 100 + 100 = 1480$$

In this instance, the HHI score of between 1,000 and 1,800 indicates a moderate degree of concentration in the market. Finally, in a third hypothetical situation, one firm commands 60 percent of the market and the balance is accounted for by two firms that each absorb 20 percent. The calculation is as follows:

$$(60^2) + (20^2) + (20^2) =$$

$$3600 + 400 + 400 = 4400$$

This hypothetical HHI score is far in excess of 1800 and is construed as indexing a high degree of ownership concentration. By similar logic, Noam adapts the HHI in other instances to calculate levels of concentration across different industries (and sub-industries) by using revenues as statistical weights. Following this logic throughout the data set, Noam calculates a score for Mass Media within the Information Sector. He also computes comparisons of concentration trends over time.

Table 1.1: Concentration in Information Sector in HHI in US (from Noam 2009: 421).

Category	1984	1988	1992	1996	2001	2005
Mass Media	564	520	580	693	1,084	1,165
Telecomm.	3,204	2,482	1,992	1,828	2,093	1,917
Info Tech.	2,638	2,077	1,724	1,823	1,926	1,879
Internet	2,959	2,114	1,685	1,287	1,357	1,494

Noam's weighted calculation of the HHI for Mass Media across all industries and sub-industries stands at 1,165 by 2005 (2009: 421). He writes that, "The average level of market concentration in the mass media is not in the range that would normally raise anti-trust action if encountered in other industries" (2005: 5). Table 1.1 indicates that, given an aggregate HHI of between 1,000 and 1,800, mass media can be deemed moderately concentrated. However, it bears mention that the HHI has been edging upward since 1988 (2009: 421). Noam points out that, although mass media still lags behind its "sister" information sector industries in terms of market concentration, it appears to be catching up due to demands for economies of scale and despite the advent of new media. Sifting further through the numbers, detailed in Table 1.2, Noam concludes that "all local mass media are highly concentrated" in the US with some differences in degree across industries (2009: 378). As concerns US newspapers as a national market, Table 1.3 shows an HHI score of 191 across more than a dozen of the biggest chains (including Gannett, New York Times Company, Hearst, and Cox).

Table 1.2: Local Concentration in HHI by Mass Media Sub-Industry in US (Adapted from Noam 2009: 373, 375, 376, 378).

Sub-Ind.	1984	1988	1992	1996	2002	2006
TV Station	2,460	2,269	2,006	1,979	1,714	1,895
Multichannel	9,344	8,960	8,836	8,529	6,300	6,629
Radio	939	1,062	1,200	2,085	2,400	2,326
Magazines	7,321	7,191	7,196	7,201	6,589	6,547

Table 1.3: Newspaper Concentration in HHI in US: National versus Local (Adapted from Noam 2009: 140, 143).

Category	1984	1988	1992	1996	2006
National Chains	155	176	200	230	191
Large Cities	5,047	5,081	4,996	5,571	5,464
Medium Cities	9,064	9,083	9,588	9,602	9,622
Small Cities	8,267	8,271	8,280	8,311	8,670
Average (all Cities)	7,219	7,239	7,367	7,612	7,676

Despite the twentieth-century intensification of chain ownership, the HHI is quite low. In contrast with European nations, including the UK's fleet of ten nationally distributed daily titles, the larger and more populous US has only three papers that could be construed as national in scope. The trio consists of *USA Today*, *The New York Times*, and *The Wall Street Journal* and, among these, only *USA Today* is geared toward mass circulation. Nonetheless, as is the pattern for the US' local media, local newspaper ownership displays striking concentration. This is because the overwhelming share of newspapers in the US do not have a local competitor; over 99 percent operate as local monopolies (Noam 2009: 142). Whereas an HHI score of 1800 crosses the threshold of high market concentration, Noam calculates a weighted HHI for local newspapers of 7,676 by 2006 across 30 large, medium and small cites in the US (2009: 143).

End of Story?

Not yet. First, the news media with which this investigation is mainly concerned is a segment of mass media for which Noam reports only preliminary findings. Nonetheless, these findings demonstrate more modest concentration with respect to news than for mass media as whole (2009: 416). Second, before interpreting the HHI scores calculated by Noam, I will also attempt to resolve an apparent controversy with respect to the very different answer to ostensibly the same question that Compaine has published. In particular, Compaine calculates an HHI score for media of 268 by 1997 as quantitative backing for his claims of low ownership concentration and vigorous competition (2004: 2). Compaine's HHI calculation is quite far from that of Noam as well as from the threshold of 1,000 that indicates moderate concentration.

Compaine obtains his low score by effectively collapsing all media into one measurement. In particular, he calculates the HHI via the top 50 media firms' overall revenues. While Compaine assumes that media convergence has irrevocably blurred distinctions between media industries, pooling across 50 firms assures a very low score as an artifact of the formula by which he elects to calculate HHI. By contrast, Noam more closely follows Department of Justice practice and groups mass media firms by industry and subindustry. In other words, Compaine's methodological decision to treat all media as one large and undifferentiated mass of economic activity determines, in advance, the outcome of analysis of the data. The opening aggregation procedure is a solvent that necessarily dilutes ownership patterns in a massive statistical solution. Baker comments that Compaine's procedures "would be merely obscurantist" if not for subsequently grounding "his argument on misapplication of traditional antitrust notions" (2007: 59).[3]

A Qualitative Take

If one accepts Noam's calculation of the HHI, the Herman & Chomsky thesis of media concentration is, strictly speaking, supported given the HHI of over 1,000. Nevertheless, the score hardly seems white-knuckle alarming since mass media is a relatively open industry by virtue of being "only" moderately concentrated. Even as Noam depends upon the HHI to gain insight into the information sector, he also wonders whether quantitative methodology is in itself sufficiently nuanced to address a matter as consequential as "media pluralism" (2009: 411).

Following a profoundly different methodological approach, Baker reframes the ownership issue by reasoning from principles about "undue power in the public sphere" (2007: 63). He asks whether mass media behaves as a widely dispersed democratic organ and expansive public sphere that is not merely informative, but educational and empowering. In this view, the stakes are high to "avoid the dangers that concentrated power poses for liberty and democracy (...) in relation to concentrated power in the public sphere" (2007: 65). Baker approvingly cites an FCC "Order and Report" from 1970 that channels the spirit of media and democracy that he champions:

> "A proper objective is the maximum diversity of ownership (...) We are of the view that 60 different licensees [to broadcast] are more desirable than 50, and even that 51 are more desirable than 50. (...) It might be the 51st licensee that would become the communication channel for a solution to a severe social crisis." (quoted in Baker 2007: 26)

Baker adds that, "useful challenges usually come from the margins" (2007: 11) and that "democracy requires diversity" (2007: 15). When news media serves its function well, the positive externalities that it generates are highly beneficial—even to the non-users of the news media. This is because the ambient dispersion of quality news on the winds of public awareness and opinion promotes better governance and public sector probity. Everyone benefits. Conversely, there are negative externalities associated with poor media performance when, for example, it scapegoats vulnerable groups or otherwise promotes "misinformation and ill-considered views" (2007: 29) that extend to the possibility of "damage to highly valued process values" within a democracy (2007: 47). Baker's theorization contrasts with Compaine's blunt defenses of the neo-liberal order that sees no ownership concentration; and, if pressed to do so, changes the topic and hails concentrated ownership as a manifestation of the merit of privileged firms, a cue for further ho-sannas to the genius of the market.

In normatively evaluating the consequences of media owner-ship, another question is begged: What is diversity? It can in a sense be served by a conglomerate with a commanding market share and many media properties. In order to cover its bets and disperse audience across its holdings, a hypothetical radio con-glomerate can pepper a local market with stations that broadcast easy listening, "oldies," heavy metal, political and sports talk (and so on). A measure of diversity will have been realized through the many "shop windows" of a single company. While bottom-line ori-ented firms may opt for a line-up of "safe" commercial products, they also possess the "'venture capital and market power (...) to launch new products'" (Hesmondalgh 2007: 76). With partial buff-ering from risk, more diversity and innovation may result. Fur-thermore, it is not necessarily correct to suppose that small media producers are inevitably tribunes of the *avant garde* and channels of the oppositional. Idealized small producers can fashion slavish imitations of the mainstream, albeit with the charm (or annoyance) of lower production values.

While acknowledging these considerations, Baker argues that the only *irreducibly* democratic form of diversity is source diversity that empanels as many different owners as possible. Dispersed ownership brings diffused authority. Source diversity through dis-persed ownership presents the structural condition that is most likely to bring about other forms of uncontrived, authentic diversity such as viewpoint diversity—and as a matter of commitment and not via a conglomerate's self-interested market segmentation algo-rithm. By contrast, mergers that reduce source diversity "could eliminate the only evangelical, the only poetic, the only leftist voice" that is available, if management sees it as unprofitable or simply awkward (Baker 2007: 70). The loss of that voice is a matter of little moment on the neoliberal cast-of-mind that, when convenient, retreats from normative judgments into the pose of dispassionate economic reductionism. In this view, the television is merely a "toaster with pictures," in the (in)famously dismissive appraisal of Reaganite FCC chair Mark Fowler (quoted in Hamilton 2004: 1). Notwithstanding Fowler's neoliberal indifference and caricatured visions of audiences staring for entire evenings at appliances, it is through culture that people make sense of their lives. And it is through news media that citizens are partly guided as they collate their ideas and experiences into the socio-political ideologies on which they act.[4]

If concentration reigns, what risks may this invite? Baker answers that "concentrated media ownership creates the possibility of an individual decision maker exercising enormous, unequal and hence undemocratic, largely unchecked, potentially irresponsible power" (2007: 16). In light of the stakes and potential disruptions to a carefully calibrated democratic order, "Even if this power were seldom if ever exercised (...) *no democracy should risk the danger*" (original emphasis, 2007: 16). Democracy should not be structurally unguarded against the mogul or firm with the size and gravitational pull to distort the alignment of political (or social, or commercial) constellations. In the light of this principle, there is no structural difference between the benign, enlightened figure and the malevolent one when too much media power has been invested in him or her. Moreover, the mogul need not seize power her- or himself (the Silvio Berlusconi/Mediaset model). He or she can stay behind the curtain and apportion consequential quantities of favor or flak toward personages, parties, programs, and proposals (the Rupert Murdoch/News Corporation model).

Conclusion: Twilight of the Moguls?

The modern gold rush of the information sector has furnished a strikingly large share of the list of the world's wealthiest persons. Among the top individuals in net worth by the 2012 rankings, Carlos Slim Helú of Mexico 21 and William Gates III lead the list having made their fortunes in telecommunications and computing software respectively (*Forbes* 2012). Other names that pepper *Forbes*' Top 50 list for net worth include information sector players Larry Ellison (Oracle), Sergey Brin and Larry Page (Google), Mark Zuckerberg (Facebook), Micheal Dell (Dell), Paul Allen and Steve Ballmer (Microsoft). However, within the media content industries, evidence suggests that the mogul is becoming a left-over of a past era, even if movement away from moguldom is slower than in the rest of the information sector. Murdoch's News Corporation notwithstanding, media firms in the new millennium are less likely to be held within a family or within a tight circle of large "insider" investors. Institutional investors (with mutual funds leading the way) and private equity are gaining ascendency (Noam 2009: 391). In this environment, there is a trend toward cold market calculation and away from the active and idiosyncratic mogul who, for good or ill, uses the media channel as a soapbox.

An array of institutions have made massive investments, salted across media industries. These institutions include State Street Global Advisers (investments of $US68 billion), Barclays ($US57 billion), Capital Research & Management ($US47 billion) and Fidelity Management & Research ($US41 billion). Consider Janus Capital's ownership stake of more than $US10 billion in Time Warner. Ted Turner's stake of $US6 billion in the same megacompany is, at the same time, larger than any other individual's holdings. Nonetheless, "hardly anyone outside the financial community has heard of Janus" while the flamboyant Turner "is legendary" (Noam 2009: 405).

What does this shift to institutional investors auger for media performance? More than nostalgia for the old days is implicated. Hesmondhalgh looks toward the economic structure when he claims that, "It is not the interests of particular individuals that are at stake but the interest of the social class to which they tend to belong" (2007: 61). The era of moguls who at times place an artisanal imprimatur on media is being supplanted by professional methods driven by a stricter adherence to market imperatives with less tolerance for "deviant trajectories" (Benson & Nevreu 2005: 6). "Tough performance mandates" demand greater fidelity to market precepts of returning investment in the short- and medium term

(Noam 2009: 397) and signify that media industries are very tightly bound to the financial interests of the investor class. Although uniquely tasked to be the information bazaar within the mass republic, media is becoming more like other businesses. It follows that news is increasingly subject to a regime more like that of any other commodity, with cost-cutting in the service of an ensemble of shareholders. It also stands to reason that cutting costs in media firms corresponds with less rigor in news gathering and production (an issue that will be explored in detail in Chapter 3).

At present, the ownership environment may feature some of the worst of both the pre- and post-millennial worlds of mass media and news. To wit, the contemporary situation presents recruitment of mass media and news into an intensified and de-personalized market regime with its attendant logics—*and at the same time* it also presents remnants of the "Age of Dinosaurs" with hypertrophic media moguls lumbering on. As aged and on-the-ropes as they each were by 2011, the Australian-born, US citizen Murdoch and Italy's Berlusconi are perhaps the two most high-profile instances that illustrate the enduring power of media moguls. The record of these surviving dinosaurs is characterized by Murdoch's transnational influence and Berlusconi's capture of media and governmental power in Italy. In Berlusconi's case, the mogul commanded approximately 90 percent of commercial television viewership prior to being elected Prime Minister (Curran 2011: 18-20). Both moguls have displayed decisive impacts in conjuring the political climate that has set the course of nations.

The cases of Murdoch and Berlusconi demonstrate that, neoliberal preoccupations notwithstanding, it is not only the State that profoundly impacts on media. The private sector and its potentates are similarly in a commanding position to steer the direction of media power and news performance. Moreover, while ownership may constitute the "loosest" of the Herman & Chomsky filters that condition news and is the most removed from what is actually reported, it is consequential as a structural factor that can and does conflict with democratic ideals through concentrations of authority. The next chapter continues to trace the arc of media performance, from ownership to manifestations of the commercial regime's impact on news media.

Notes

[1] Milton Friedman (1912-2006) was perhaps the twentieth century's leading neoliberal academic—and a perversion of the public intellectual for his complicity in a criminally abusive right-wing dictatorship. See Hitchens on the gangsterism of Pinochet (2001: 55-89) and Friedman's blandly technocratic defense of his roles as advisor to the regime (Letelier 1976). As neoliberalism's critics have argued, the vicious class war that its policy package unleashes demands that subject populations are shocked and traumatized into tense acquiescence via, for example, coups, disasters (real or contrived), and police state conditions (Klein 2007).

[2] Advocacy for radio behemoths stands in contrast to their behavior when confronted with competition from micro-broadcast "Low Power FM" (LPFM) radio stations. Under the tendentious rationale of alleged "spectrum interference", the commercial industry moved assertively against Clinton-era efforts to recognize LPFM (Hoynes 2002; McChesney 2004). The episode is illuminating in the light of neoliberal insistence that free speech, localism, anti-elitism, democracy (*etcetera*) coincide harmoniously with unbridled corporate hypertrophy.

[3] If Compaine stacks the deck in assembling data, he is not alone. Under the chairmanship of Michael Powell (2001-2005), the FCC had little regard for even measuring corporate hypertrophy. The FCC's Republican Party majority proposed a "Diversity Index" (DI) that was designed to look like the HHI (scores up to 10,000, threshold of concentration at 1,000), even while its assumptions were starkly distinct. Along with an absurdly expansive concept of what constituted "local" media in a metropolitan area, each media outlet was assigned the same statistical weighting as a "voice" *irrespective of the size of its audience and revenue.* Under this formula, the television station at Duchess County Community College, with an audience of dozens, was construed as a "voice" equal to that of New York's ABC affiliate. Baker claims that the index "borders on the fraudulent" (2007: 79) in the light of assumptions that program low concentration scores into the DNA of the calculations (also see Noam 2009: 371-72).

[4] In a different register of diversity, it is also apparent that market mechanisms have failed in realizing it. While "minorities constitute" 34-percent of the US population, they own 3 percent of the full power commercial TV stations. Women clock in at 6 percent on the same ownership measurement (Copps 2011: 294).

CHAPTER 2
News in a Neoliberal Milieu

The challenge of American newspapers "is not to stay in business—it is to stay in journalism"

—*Sunday Times* (London) editor Harold Evans (quoted in Bagdikian 1992: 137)

It is part of the informal "script" for the endgame of modern revolutions against abusive State power to capture the regime's media organs. In the case of Libya's revolt against Colonel Moamar Gadafi, the rebels decided to pull the plug on the official Jamahiriya station as they stormed Tripoli in August 2011 (Halliday 2011). Abusive regimes' often meet this closing scene because, as C. Edwin Baker notes, they "regularly try to censor or control the mass media's provision of vision and information." It follows as a corollary that "The health of democracies (...) depends on having a free press" as an essential national schoolroom (Baker 2007: 5).

In discussing constraints on media, it is less commonplace to shift the locus of concern from heavy-handed governments to the scale of the private sector's influence. Nonetheless, some scholars have made this move. W. Lance Bennett argues that the US has constructed "a national information fortress with one wall, leaving the information system completely vulnerable to predation and degradation at the hands of little controlled business interest" (2001: 12). Baker comments in a similar vein that private sector influence constitutes "a social control mechanism often more effective than government censorship in limiting press freedom" (1994: 56). The Herman & Chomsky model concurs in positing, via its second filter, significant distortion and inhibition of journalistic behavior within the advertising-driven milieu—and in the absence of a censorious State.

Currently, US news media (the press, in particular) is enveloped in commercial upheaval. For example, *The Los Angeles Times* slashed its workforce by half and then its parent, Tribune Company, declared bankruptcy. A number of major papers—*Baltimore Sun*, *Boston Globe*, *Philadelphia Inquirer*—have closed their remaining foreign bureaus even as globalization is galloping forward (Starr 2011). Not only foreign bureaus have been shuttered. The Cox Newspapers chain that formerly stationed 30 reporters in Washington now has none. The effect trickles down further. Starr (2011) observes that local statehouses are now increasingly bereft of beat reporters.

In more general terms, Davis identifies the structural factors that are battering the journalistic enterprise in the English-speaking world. These factors include "declining employment security and job cuts, the hiring of cheaper, junior staff replacements, decreasing editorial resources, an increase in output and paper supplements, efficiency drives, the growing power of accountants within firms and a greater dependency on externally supplied 'information subsidies'" (2010: 129). The collapse and reconfiguration that is widely agreed to be occurring in the news industry has occasioned alarm (Gitlin 2011), profoundly mixed feelings (Hedges 2011), and some mutterings to the effect of "good riddance" (Jackson 2011).

Cooper's eulogy for the now passing "old model" of newspapers chimes at unsentimental bells as it describes the "why" behind its collapse. He fingers a strong degree of complacency and shortsightedness:

> Excused from competition, the industry was less responsive to consumer demand (...) [It] failed to innovate, and dissipated economic rents in inefficient and socially unproductive ways. The monopoly and oligopoly situation created perverse incentives to squeeze (out more) profits by cutting quality rather than investing in productivity. (Cooper 2011: 327)

Simon is blunter as he scorns management's model during the 1990s that was devised, he notes, by managers trained in business and not journalism: "news was the stuff trowled into the columns next to the display ads" since there was believed to be "more profit producing a half-assed, mediocre paper than a good one" (2011: 49). In other words, supposedly business-savvy management was producing an unsatisfactory product and failed to retool the news factory when it had its chance. With readership subsequently shifting toward the internet with its far stingier advertising rates, revenue has crashed by about one half for the newspaper industry (Cooper 2011: 336).

At the same time, amidst the sky darkened with locusts, the market-grounded crisis is strangely chimerical. Consider that 40 percent of people in the US still claim to read a newspaper (i.e., tens of millions of people). *The Houston Chronicle* generates close to a *billion* internet page views per year, and it does so with a staff of 206 that is half of what the paper employed in 2006 (Grueskin, Seave, and Graves 2011: 21). Flagship "papers of quality" such as *The New York Times* "are more widely read than ever" due to internet distribution (Starr 2011: 24). *The Times* claims a hardcopy weekday circulation of 900,000—and internet readership of 30 million. Here, at the hinge between old and new media, is where some of the current conundrums arise. To wit, the *Times*' hard copy readership constitutes only 3 percent of overall readership with the advent of the internet. Hard copies, nevertheless, generate 80 percent of the paper's revenue (Grueskin, Seave, and Graves 2011: 21).

While change is in motion, Freedman places it in this perspective: "The internet does present a genuine, if overhyped, challenge to the business operations of traditional news organizations" (2010: 50). Despite dramatically lower ad revenues drawn from the internet as a "cost per thousand" readers, newspaper profitability in the recession year of 2008 was at an industry-wide average of 11.5 percent. This figure was down from an average newspapers profit margin of 19 percent in 2006 when the internet was believed to have already poached far into the print media's business model. By contrast, the 500 largest industrial firms lurch forward with a comparatively modest 6.8 percent average profit margin (Baker 2007: 32-33).

Whatever the upheaval of a particular moment, on a structuralist view, the demands of commercialism shape US news performance in a pervasive, everyday fashion. In Bennett's appraisal,

"Business-driven news formulas dictate manufacturing the most dramatic audience-grabbing stories for the least cost and with a minimum of attention distracting complexity" (2001: 19). Moreover, "hard news" ("what an informed person should know" [Bennett 2001: 12]) is literally a tougher sell within a more commercialized environment, and so less effort will be devoted to it. A cycle begins. Less effort is devoted toward cultivating the taste for probing journalism over the ratings crack of "soft focus" and sensationalism—thereby further dampening demand for hard news. With commercial prerogatives welded into place, hard news on television receives less time than commercials while news is often selected by "marketing and demographic guidelines" in order to better deliver an audience that appeals to "particular kinds of advertisers" (Bennett 2001 :12).

Having introduced the chimerical crisis of news as stressed but still highly profitable, the balance of this chapter will elaborate on the economic pillars that have long supported journalism. The account presented here squares with that of Herman & Chomsky in its emphasis on commercialism and the implications of advertising as the main revenue stream for media enterprises. However, I unpack these issues more extensively than Herman & Chomsky do in introducing their model. In doing so, I follow Baker's carefully calibrated theorizations on media economics. Baker posits a place for advertising revenue in subsidizing media (1994: 83). However, he is adamant that private sector influence (like State influence) should be carefully tempered by a robust structure of laws. As matters currently stand, Baker opines that, "private entities in general and advertisers in particular constitute the most consistent and the most pernicious 'censors' of media" (1994: 3) in part for the degree to which media has been absorbed into the regime of big business.

More specifically, I will argue that news media's grounding in a commercial model begins with advertising as the main source of revenue. The dependence on ads has enabled the strategy of capturing the audience segments with more desirable demographics—typically, more disposable income—through content that appeals to such niches. At the same time, the relationship between commerce and news is complexified by upsurges of professional independence.

News media can and occasionally does bite the commercial hand that feeds it. "States of exception" from the rigors of the commercially driven writ of objectivity also exist, via management's political interventions that are typically enacted from a right-wing angle. The chapter finishes with an economically grounded analysis of the new media that postdates Herman & Chomsky's original model. Nonetheless, discussion of the economics of new media demonstrates that Herman & Chomsky's emphasis on the commercial base of the industry yields enduring insight into the bases of journalistic behavior.

The Class Dimension

News leads a double-life in the US. Firstly, it is the nourishment that sustains Enlightenment-style idealized democracy. Secondly, news is yet another commodity that is bought and sold. Regarding the second moment of this double life, hard copies of US newspapers are commodities that for centuries have been subsidized by two main sources. The first is the reader who buys a copy and pays either at the newsstand or via subscription. The second, less obvious source of subsidy arises from the promotional material on the newsprint. Under this arrangement, the newspaper is doubly commercialized. That is, a paper sells its content to the readers. It then pivots around and sells its readers' attention to the advertisers.

As the situation has developed over time, the standing of the two constituencies that furnish income is by no means symmetrical. While readers would seem to be the "natural" first constituency of the newspaper, when one follows the money, it is not even close. Newspapers obtain about 75 percent of their revenue from ad accounts and devote about 65 percent of newsprint space to them (Baker 1994: 7). Moreover, commercial broadcast (network TV, radio) is almost totally beholden to ad accounts for revenue. Beyond the specific numbers, it is trivial and obvious that some entity—whether it is the paying audience, advertisers, State subsidizers, or trusts—must furnish the revenue stream for media. Why does it matter if advertising is the biggest source? Regardless of who funds it, the news cannot be fabricated from whole cloth since events intervene in the first instance to ground the content of news accounts. Moreover, all sources of revenue have their associated compromises with respect to demands that benefactors may make.

At issue is the impact of the private revenue stream that is both pervasive and perhaps particularly subtle. US news media is not merely proximal to business *but is itself* a division within the private sector. Media and journalism's recruitment into the regime of business and commodification places them on a market footing and conditions their sensitivities toward the interests of the private sector of which they are a part. Extending this view, the neoliberal "free market" has been construed as "more effective than state controls in conscripting the press to the social order" (Curran 2011: 148). High-profile journalist Richard Wolffe corroborates, albeit in defense of wheeler-dealer entrepreneurship under the rubric of being a press man:

> The idea that journalists are somehow not engaged in corporate activities is not really in touch with what's going on. Every conversation with journalists is about business models and advertisers (...) You tell me where the line is between business and journalism. (quoted in Greenwald 2009: 4)

More specifically, what implications follow when the advertisers that buy the audience's attention constitute the first constituency? While racking up large circulation is important, in many circumstances attracting the optimal demographic is emphasized even more. The most desirable demographic is endowed with greater disposable income and the associated business model prioritizes news of interest to a wealthier audience. News about chic barrios and vacation destinations, distinction-endowing new gadgets, and investment opportunities are *de rigueur*, and not articles about how to acquire food stamps. By the 1980s, editors had been trained to sniff out news of interest to wealthier readers that would seamlessly stitch the advertising and news content together (Bagdikian 1992: 232).

Within this environment in which demographics matter, the news organ seeks advertising accounts for bigger ticket items. Displaying autos and designer clothes constructs a "shop window" that is more lucrative than ad accounts that promote gum and softdrinks. Elaborate demographic reports on the readership are designed to convince advertisers that a media outlet hails an audience with suitable disposable income. By the 1970s, *New Yorker* marshaled 134 pages of reader demographics to potential advertisers (Bagdikian 1992: 117). Publications can and do attempt to raise circulation. However, they may do so with an interest toward not raising circulation so high as to water down reader demographics. On this market-driven logic, publications have even attempted to actively shed readers and strip down to an exclusive, up-market demographic profile. The strategy was recently pursued by *Newsweek* during its (ultimately ill-fated) "re-branding" episode (Kishtwari 2009).

More recently, the Project for Excellence in Journalism (PEJ) compiled figures on whose interests are not likely to be covered in the news. PEJ's five-year study found that 0.1 percent of news reports concerned the poor (32 out of almost 24,000 news stories studied). A similarly stingy 0.2 percent of stories were devoted to the elderly (57 stories out of 24,000). By contrast, there were more than 500 stories on celebrities in the sample. While reports on celebrities do not dominate the news, they compose more than 2 percent of stories and trigger more than five times as many news items as the more ubiquitous poor and elderly combined (Project for Excellence in Journalism 2003: 16). Although there is surely more than one reason for marginalized groups' near invisibility, lack of disposable income figures as a factor in banishment from news.

Within its particular media environment, James Curran observes similar tendencies in Britain. He writes that, by the nineteenth century, "Papers were forced to close down with circulations far larger than their rivals" (2011: 151). In the case of the *Daily Herald*, its circulation of 1.265 million was *five times* the surviving *Times* when it closed in 1964 (2011: 165). Ad revenue deficiency was the cause of death. *The Times* attempted to curb subsequent readership gains to avoid dilution of its up-market demographics. Numbers aside, Curran concludes that well-heeled elites have been "served by papers that briefed them effectively as citizens" and have "offered extensive coverage of public affairs structured in terms of their concerns, agendas, and interest" (2011: 166). It is one more illustration that, in class-divided societies, socioeconomic standing impacts on everything from food (Patel 2007), to sport (Bourdieu 1999), to life expectancy (Wilkinson 2005).

When Commercialism Reigns, What about Content?

In discussing his ethnographic findings from the 1960s and 1970s, Gans maintains that "top producers and editors would not consider killing a story because it might antagonize advertisers" (2004: 253). A firewall divided content from commerce. However, under the sway of the neoliberal regime installed at the end of the 1970s, the pull of market interests has intensified. By the 1990s, Baker elaborates on what is and what is not in the news when ad subsidy and private sector prerogatives reign:

> Don't like the news that connects the product you sell or your company to death or murder? Demand silence. Want an appealing, upbeat media environment for your ads? Pay to get it. Concerned that your advertising expenditures are wasted on media consumers, like the poor, who are unlikely customers? Just tell the media producers to stop providing material of interest to that audience. (1994: 44)

Does this sound exaggerated? Consider that, by the late 1970s, two of the most pervasive causes of premature death in the US were alcohol and cigarettes. Simultaneously, these two industries were also the two biggest advertisers in magazines (Baker 1994: 5). By 1980, however, the US press published more reports about the causes of polio and tuberculosis—scourges that had been effectively eradicated by then—than it did about the cause of one in seven deaths in the US from cancer (Bagdikian 1992: 170). At the same time, *Reader's Digest* had been subject to ad boycott from the cigarette industry for decades following its iconoclastic, critical accounts of smoke-related health hazards during the 1950s (1992: 172). The *Reader's Digest* episode is one among many instances of the carbon monoxide industry's multi-faceted push-back at critical examination (see Oreskes & Conway 2010: 10-35).

A striking display of advertiser power occurred during the first "Gulf War" of 1991 when advertisers were anxious for their products to be placed within a positive environment. Hence, they were concerned that any hint of bad news from the front would furnish unwanted mood music before the cut to a commercial: "'Sponsors don't want to shift from a general discussing K.I.A.'s—soldiers killed in action—to "Double your pleasure/double your fun." It's jarring."' (advertising executive, quoted in Carter 1991: 1). Rather than prompting defiance from the TV networks, management scrambled to allay advertiser anxieties through promises to "insert the commercials after segments that were *specially produced with upbeat images or messages about the war*, like patriotic views from the home front" (emphasis added, Carter 1991: 1). These contortions to re-script the news as feel good material were enacted while the networks were ostensibly covering one of the decade's biggest (literally life and death) news stories.

In a recent study, journalists "report more cases of advertisers and owners breaching the independence of the newsroom" (Kovach, Rosenstiel, & Mitchell 2004: 27). One-third of reporters have indicated such pressures occur at work in a Pew Research Center study that surveyed more than 500 news workers. Moreover, substantial majorities of news workers agree that "increased bottom line pressure is seriously hurting the quality of news coverage"; 66 percent of national news workers and 57 percent of local news workers concur (Kovach, Rosenstiel, & Mitchell 2004: 28).

Deep immersion in commercialism brings further implications. When a firm is engaged in more industries, it follows that there are more opportunities for critical reporting to negatively impact on one of its subsidiaries. Beyond external relations with other large firms, "sibling conflicts" between different divisions of the same media company have also developed. In anticipation of such cases, Time, Inc.'s managing editors sign an agreement to "'not at any time denigrate, ridicule, or intentionally criticize the Company or any of its subsidiaries or affiliates, or any of their respective products'" (*Time* magazine's editor-in-chief, quoted in Baker 2007: 40). News organizations have indeed reported on disorders within other divisions of a corporate empire. For example, during the summer of 2011, News Corporation's Fox News reported the 13 arrests (as of 18 August 2011 [Reuters 2011]) made at News Corporation's British subsidiary in the opening thrusts against phone hacking accusations. However, at the moment that the explosive story was gaining legs elsewhere, News Corporation's Fox News subsidiary moved sluggishly toward it and with uncharacteristically muted commentary as compared with its usual carnivelesque offerings (Rothstein 2011). Questions also arise around the issue of how to finesse coverage that may impact a media firm's commercial relations with other firms (Greenwald 2009).

Counter-examples to the commercial writ and "heroic professional stands" also exist (Baker 2007: 41). Journalists and editors may indeed revel in opportunities for displays of independence. The fact remains that the structural form of media industries presents some built-in obstacles to going where the truth may lead. In colorful language, Baker claims that ad revenue has power that in its most raw form "prostitutes broadcasting to its paymaster" of the corporate order and ad accounts (1994: 110). This is, in fact, obvious—and would be readily recognized as such if it was Statist/commissar culture that demanded such finesse from news media toward its powerful benefactors.

Neoliberal Reveries

By contrast with cautions on commercialism from critical scholars, neoliberals exhibit complacent postures toward corporate-produced news media. In the idealized neoliberal view, private sector news is akin to a transparent pane of glass through which one gazes on the world. Only deviations from professional practices complicate this scene in which the world largely is as it is reported to be by private media. Benjamin Compaine channels this view: "Other than an anecdotal story here and there, there is no suggestion that the managers of major media companies are individually or in concert fostering a political ideology or suppressing an ideology through the media properties they program" (2005: 46). In a similarly neoliberal vein, James Murdoch of News Corp chimes in with this aphorism: "The only reliable, durable, and perpetual guarantor of (journalistic) independence is profit" (2009: 19). In this view, an aberrational reporter or editor "going rogue" presents the lone disruption to systematic, high-fidelity transmission of reality via private sector news media.

However, like a volcano that lies dormant for extended periods, even activist ownership and management is not perpetually interventionist. Rupert Murdoch's steering of News Corporation furnishes an example of interventionism tempered by some discretion

(Stecklow, Patrick, Peers & Higgins 2007). Nonetheless, one reporter for the London *Times* claims to have had 200 stories that he composed on the People's Republic of China (PRC) spiked for criticisms that would endanger the company's prospects for "emerging satellite, cable, and Internet markets" in the Asian giant (Bennett 2001: 161). News Corporation management also reached down to spike former Hong Kong governor Chris Patton's book contract with its subsidiary HarperCollins on similar grounds of not affronting the PRC. Indignantly fashioned neoliberal aphorisms notwithstanding, independence within the discourse produced by the Murdochs' own firm is manifestly vulnerable.

Furthermore, commercial ideology has long been deeply embedded in the structure of US society (Boorstin 1992; Ewen 1976; Packard 2007). It is reasonable to infer that private sector media organizations (their workers, management and ownership) are *always already* subject to pervasive demands from commercial ideology. In this respect, there is typically little need for management "to individually or in concert" promote or squelch an ideology with respect to their workforces, anymore than one would need to compel Bostonians to favor the Red Sox baseball club. This is because the primacy of commercial ideology is already understood, internalized and in the main eagerly reproduced as it furnishes the framework in which the news organization operates.

Bagdikian (1992: 27-89) and Baker (1994: 46-56) cite a litany of management decisions that have been shaped by commercial ideology. These decisions include proscriptions of criticism of business that did not bear on any immediate interest of a media firm, but that could implicate the "mood music" of capitalism. Nonetheless, the degree of professional integrity internalized by news workers insures that courageous stands are taken and that commercialism is not mechanically reproduced even as it is highly privileged. In one notable case, London *Times* editor Harold Evans campaigned in the early 1970s on behalf of victims of thalidomide. The campaign cost the paper its biggest ad account of 600,000 pounds per year as the parent company behind the drug withdrew all ads from the paper. In taking this costly stand, Evans nonetheless enjoyed the backing of management. Editors are not always so fortunate and have suffered stinging professional setbacks for confronting commercial interests, as Meech (2008: 39) documents in the passage that follows his account of Evans' principled stand.

While management can and does reach down into news content for business and/or ideological reasons, it is not straightforward to predict when or how this will occur. Moreover, due to pride, professional integrity, and a sense of service, even devotedly pro-capitalist media workers push back against commercial pressures in at least some instances. In this vein, through surveys of US news workers, Kovach, Rosenstiel & Mitchell report that friction about commercial pressures "seem to have widened the divide between the people who cover the news and the business executives they work for" (2004: 27). However, commercial calculus enjoys crucial enablement when, under the sway of neoliberalism, the regulatory posture that State agencies adopt is one of non-regulation; in Britain, so-called "light-touch regulation" is enacted via the notably passive Press Complaints Commission (Phillips, Couldry & Freedman 2010: 59). In light of the evidence, Compaine and Murdoch assert ideological fantasies, not a description of reality, when they repeat the simplistic meme that private sector media management must be behaving with dispassion toward commercial implications of the news content it produces

State of Exception

Compaine insists that, "Profit not ideology" rules the mediascape and political views are routinely hushed in favor of the hum of the cash register (2005: 39). Some leftists, such as David Reiff (1993), similarly hold that capitalists will eagerly sell Mao caps and symbolically trash their own ostensible ideology if it stimulates sales. In this view, it is "All about the Benjamins" and the monetary bottom line. Putting aside the degree to which the bottom line orientation is indeed ideological in itself, there are also significant exceptions to the alleged reign of commercial benefit when politics bulldozes into the news.

More specifically, the US' news ensemble tends to hover around the political center through the generally prevailing doctrine of objectivity. This concept of objectivity is nourished by support of business imperatives, deep class striation, and the US' international adventurism in "defense" of the resultant political center and its version of the status quo. There are, nevertheless, moments when so-called centrist business calculation is cast aside to reveal the "state of exception" in which ideological preferences are nakedly indulged. Given the economic wealth of media firms and their management, these states of exception in which ideology is rammed into the news hole tend to have a distinctly right-wing odor.

Consider the case of left-of-center US television host Phil Donahue who was fired from his MSNBC program in the spring of 2003. Curiously, Donahue was jettisoned while his program sported the network's highest ratings. Donahue's political stance—and *not* the bottom line concern of advertising revenue—was fingered by management as the reason for dismissal in a leaked internal memo. The host was judged a "'difficult face for NBC in a time of war,'" as "'he seems to delight in presenting guests who are anti-war, anti-Bush and skeptical of the administration's motives'" (quoted in Greenwald 2008: 3). The memo notices that "'our competitors are waving the flag at every opportunity.'" In other words, MSNBC was missing a chance for commercial differentiation by waving the same flag *away* from the incipient and disastrous invasion and airing thoughtfully patriotic criticisms of neo-conservative dogma. One would expect that such differentiation and its market-grounded logic would be considered by an independent news organization as a matter of course.

Donahue recounts that management compelled him to schedule more pro-administration voices on his program by a ratio of at least 2:1 in the run-up to the invasion. On one of the defining issues of the first decade of the millennium, management demanded no contrarian (anti-administration) guests on the program unless outnumbered (put on defense). Reporter Jessica Yellin further corroborates the pro-Bush administration demands from MSNBC management, on what is today known as the "liberal" news network, mythologized as symmetrical to News Corporation's rightist Fox News.[1] In an interview, Yellin states that

> the press corps was under enormous pressure from corporate executives, frankly, to make sure that this was a war that was presented in a way that was consistent with the patriotic fever in the nation and the president's high approval ratings. And my own experience at the White House

> was that, the higher the president's approval ratings, the more pressure I had from news executives (...) to put on positive stories about the president. (quoted in Greenwald 2008: 1)

Beyond the pressures that Yellin describes, the state of exception from the commerce-friendly writ of objectivity may begin with the decision to continue to produce publications that are longtime money losers. This occurs presumably because their owners regard these business failures as ideologically serviceable. Examples include self-proclaimed "Emperor of the Universe" Sun Myung Moon's *Washington Times* (see Brock 2002) and the Murdoch/News Corporation cash sink, *The New York Post*, both of which are resolutely right-wing papers. As concerns the latter, *Businessweek* marvels that, "The *Post* has lost so much money for so long that it would have folded years ago if News Corporation applied the same profit-making rigor to the tabloid as it does to its other businesses" (2005: 1-2). Furthermore, if keeping the cash registers ringing was alpha and omega, the troubled figure of Glenn Beck (Zaitchik 2009) would have been removed from Fox News' line-up far more rapidly than he was for being demonstrably radioactive to advertisers. By early 2010, over 200 Fox advertising accounts were actively avoiding association with him (Kurtz 2010h). Fox finally jettisoned Beck in mid-2011, long after it was apparent that he was a notable cash-flow suppressant.

The reign of markets demonstrably does not vet ideology through a dispassionate cash-generating calculus. Notwithstanding Fox' coy tagline of being "fair and balanced," critiques have long been made about the station's habit of thumbing its nose at even pretenses of either (Ackerman 2001). What is apparent looking in from the outside at News Corporation's Fox is also evident from the inside. Former Fox producer/editor/writer Charles Reina reports that the adherence to the admirable if flawed doctrine of objectivity that he had observed across a long career at other news outlets was inoperative at Fox (Grieve 2003). In Reina's account, management distributed daily memos suggesting story angles favorable to the Bush administration and Republican party. Out of fear, or for being "true believers," Reina reports that news workers delivered what management was expecting. More recently, Fox campaigned on the air to raise money for Republican Senate candidate Scott

Brown in a 2010 Massachusetts special election to succeed the deceased Edward M. "Ted" Kennedy (Boehlert 2010). These actions brazenly defied the regulatory distinction between news and advocacy for a political party although, in a deregulated neoliberal environment, no penalty came of it. Moreover, News Corporation is not alone in behavior that is troubling toward neoliberal assertions about generalized ideological indifference in private sector news, as the case of Sinclair Broadcast Group further demonstrates (Lieberman 2004; Sourcewatch n.d.).

Entire networks devoted to grinding an ideological (right-wing) ax contradict Compaine's dismissal of ideological diktats from management as "an anecdotal story here and there" (2005: 46). In summary, business and its very tangible interests pervasively shape the news that emerges from the private sector firms that produce it. Moreover, while far less prevalent, contemporary management can and does put commercial interests to the side to bulldoze into the news with dedicated right-wing ideologizing.

Cyber-Saviors?

A key issue to examine in bringing the Propaganda Model into the present is whether the internet presents significant pushback against the myriad defects of incumbent media that Herman & Chomsky identify. Moreover, how have internet economics impacted on the prevailing commercial model?

In assessing the terrain of new media scholarship, Natalie Fenton observes (2010a) that the internet has provoked binarized discourses about its impact. It has become *de rigueur* to exalt the internet as the one-stop solution to entrenched problems associated with the old incumbent media order (also called "heritage" or "legacy" media). It has also become a commonplace to posit that the internet has no more than illusory possibilities for betterment in the news ecology. Fenton theorizes that neither internet evangelists nor doubters hit the mark. In a carefully calibrated interpretation, Fenton characterizes internet evangelists as positing technology as possessing its own life force (techno-essentialism) that elides social realities on the ground. Alternatively, the internet doubters suffer a failure of imagination by visioning the internet's present and future as an inevitable replay of the old media past. For the moment,

I seek to present an economic explanation for why the internet has not yet realized its potential to rollback incumbent media.

The internet is not simply an elaborate echo of what has come before it: "As a repository of information and knowledge, it is unparalleled" (Fenton 2010a: 14). Moreover, its substrate in digitalization makes for a "common currency" of convergence among the previous waves of incumbent media. Several other features have occasioned optimism about the internet's impact on the mediascaspe. These include what appear, at first blush, to be very low economic barriers into the media arena. Besides the modest cost of production and distribution, internet content is unfettered from the geographic limitations that have long bound human interactions and limited local enterprises (Hesmondhalgh 2007: 248). Between the text and readers, there are also substantial opportunities for feedback and collaboration. In its brief history, the internet has provided infrastructure for bottom-up organization of counter-hegemonic actions.

Despite its high ceiling potential, here is a "killer fact" that speaks to the difficulties that the internet presents for existing business models. Noam reports that "a print reader (of a newspaper) generates *more than 20 times* as much in revenue than an online reader" (emphasis added, 2009: 440). Do the math and further implications follow. Specifically, if a newspaper loses even 5 percent of its print audience, it would need to *double* (increase by 100 percent) its readership through recruitment of new online readers in order to restore the lost revenue. The slippage is accounted for by the fact that internet ads do not bring in the same level of advertising revenues. As of 2007, only 7.5 percent of US newspaper ad revenue was generated by the online versions of papers despite their eye-popping number of visitors and page views (Freedman 2010: 45). In comparison with a newspaper hard copy, readers tend to spend less time with a webpage and look at fewer articles. This pattern of "grazing" limits the ad rates that internet content commands (Freedman 2010; Grueskin, Seave, and Graves 2011). Furthermore, there is a glut of internet content on which to advertise that also suppresses the revenue generated by any given advertisement.

Now, step back to the big picture. Privatization and commercial imperatives have made news media dependent upon advertising revenue. These pressures have enabled the attendant compromises and systemic blind spots for journalism as sentinel of the public interest. As ad revenues shrink, one can see that news media has long been at once *exploited by* and *dependent upon* its private benefactors. The paradox follows that, while the internet and overall readership has boomed, newsrooms shriveled and advertising revenues nosedived.

At the same time, while the still prosperous incumbent media's revenue model has been shaken by the internet, it retains the capacity to retool on the fly. Emerging new media outlets do not have this same luxury. They deal with limitations endemic to new media substrates, such as attenuated revenue, even as they simultaneously assay to challenge the better positioned old guard. In a grimly hilarious turn, one of the ventures held up as an internet success story in Downie, Jr. & Schudson (2009)'s largely upbeat report on internet enterprise "talks back" at them. In the online report's comment section and in business savvy jargon, Leder writes,

> While it's nice to see footnoted.org mentioned in the report, the model for sites like mine is still pretty sketchy. (...) Although trade publication advertising is well-established offline, it's very difficult for small sites like mine to tap into that market without hiring sales people, which leads to expensive overhead. $2 and $3 CPMs [cost per thousand] and Google Ads barely cover my monthly VPS [Virtual Private Server] bill and there's lots of studies that show that more sophisticated readers rarely click on ads. While attention from big media outlets has been nice in helping to build the brand and credibility of the reporting, it doesn't exactly pay the bills. (Leder 2009: 27)

Although claimed as a new media triumph by Downie & Schudson, Leder's testimony demonstrates how tight the current economic environment is for online organizations. The tenuousness is even greater when one considers that many of these fledgling micromedia efforts "are at risk of being destroyed by a single lawsuit" should they step on the wrong toes through rigorous investigation or otherwise attract flak (Starr 2011: 35).

The internet is, moreover, subject to the "same dynamics prevailing in the traditional media leading to [ownership] concentration" (Noam 2009: 424). In light of its moderately high concentration figures in the Herfindahl-Hirschman Index (HHI), Noam cautions that the internet will not by the mere fact of its existence annul prior concentration issues. Indeed, the internet may be entrenching the incumbent-media "dinosaurs" that are supposedly on the brink of extinction. Seventeen of the 25 most visited online news sites are organs of incumbent news firms. The four most visited sites (CNN, Yahoo!, MSNBC, AOL) account for more than half of all news-seeking by online readers (Noam 2009: 424). Moreover, an estimated 95 percent of "original reporting" that appears on a swath of websites can trace its origins to incumbent media (Copps 2011: 290).

A similar situation, in which the internet piggy-backs onto incumbent media, is evident in Europe. For example, the roster of Britain's top ten news websites is composed of seven incumbent media outlets (BBC News, Sky, *Guardian*, *Times*, *Daily Telegraph*, *Daily Mail*, and *Sun*) augmented by three news aggregators. While the aggregators are relatively new firms (MSN, Yahoo! Google News), they do not engage extensively in original journalism (Curran 2011: 115). Aggregators' selections of articles also tend to reify and reinforce the news judgments of established players. In this vein, no alternative news media were returned on the first page of a Google search in one study, which means they are often effectively invisible (Redden & Witschge 2010: 180). Between the evidence of increased burdens on journalists to produce and the rise of news aggregation sites, Fenton concludes that, "Far from breeding a diversity of views, online news content is largely homogenous" (2010a: 15).

As the present is never a carbon-copy of the past—and it is politically debilitating to assume that it is—one can expect breakthroughs with respect to quality internet media. Through its tumultuous history since "going live" in 1995, *Salon*.com demonstrates that internet enterprises can enduringly succeed while providing original content (Goss 2003). Nonetheless, despite some noteworthy contributions toward improved news ecology, the cyber universe is mainly duplicating established concentration patterns at present. For the foreseeable future, the very narrow profit margins that loom over on-line journalism present a structural constraint.

Non-Commercial Configurations

This chapter has engaged with the criticisms of media commercialism and its impact on journalism. Beyond new media, what alternatives exist? One British paper touts its partial independence from the private sector while making a sales pitch for subscriptions. *The Guardian* writes that it "is the only daily British newspaper to be owned by a trust. All proceeds from the Guardian Media Group, of which the Scott Trust is the only shareholder, are reinvested in our journalism." As *The Guardian* carries numerous advertisements, its separation from the private sector must be understood as relative and not absolute. Nevertheless, *The Guardian* takes jabs at Murdoch's News Corporation that it had been investigating over illegal phone hacking. *The Guardian* writes, "That independence—and the power it gives us to do sustained investigative journalism—is increasingly rare in a media world dominated by global media companies and billionaire proprietors" (*The Guardian* 2011a: 2). *The Guardian* has also been the most innovative British paper with respect to moves online. Even before it was the industry standard, the paper published articles on the *Guardian Unlimited* website before a story went to newsprint. The *Guardian Unlimited* website also attracts the largest internet audience of all incumbent British newspaper titles (Franklin 2008: 1-2, 24-27).

In the US, the non-profit, Poynter Institute owned *St. Petersburg* (Florida) *Times* is the exception to the rule of private sector ownership of the news. However, a non-profit trust for every paper is not likely to be realized. In the US, scholars have envisioned an array of solutions to commercial pressures. Proposals range from technocratic tweaking (Hamilton 2004: 5-6), to reformism grounded in further market discipline (Grueskin, Seave, and Graves 2011), to some tentatively formed non-market models (Downie & Schudson 2009). Rodney Benson (2011) cuts straight to a solution. He writes, "In a system with multiple types of ownership and funding—private, government (with guarantees of independence from direct partisan control), non-profit, journalist-owned (as at *Le Monde),* etc.—there is a greater likelihood of ensuring that no powerful actors or public problems will be able to elude critical journalistic attention" (2011: 198). In seeding a more expansive media ecosystem, with vigorous private and public media sectors, subsidy from the State may assume two forms. Subsidy may be allocated toward what are mainly private enterprises to further support their continuance. Alternatively, a media organ may be a State-funded corporation in its own right, operated on a service model.

As Hesmondhalgh argues, the public service broadcast option has enabled "some dubious monopolies." Moreover, it is not desirable to supplant private media domination with a State-funded version of the same. However, when backed with a careful blueprint, the State has constructed news organizations that have practiced the ideals of "the (public's) right to access to high quality programming of all kinds" (Hesmondhalgh 2007: 269). Scrupulously fashioned public broadcast promotes diversity alongside the compelling interest in fostering republican solidarity within shared mediated space (Curran 2011). Given its potential power, it is crucial that public broadcast be mandated and monitored for inclusivity and fairness in serving the entire nation and its various sub-cultures. Management must also, by statute, be substantially independent from and not subject to whims of political administrations; that is, public broadcasting requires insulation from flak-driven "political scrutiny" as well as "retaliation for critical programming" (Baker 1994: 113). While these are demanding standards, Curran (2011) observes that the BBC is buffered from the scourges of both market and government censorship. The BBC is also supported by an ample funding stream while simultaneously subject to a governing trust, independent advisory panels, a parliamentary select committee and regulatory scrutiny. This is far more oversight to assure independence from benefactors than private media experiences in the US (Copps 2011). The resultant BBC is one of the most widely known and most highly regarded projections of Britain, both within and beyond its borders, via the attractant of quality media.

Hackles are predictably raised by neoliberal dogmaticists over any social democratic governmental activity.[2] Concerning the First Amendment, Baker argues that the reflexive appeal to it is frequently misguided or misleading. The amendment states, in negative terms, what the government *cannot* do to suppress journalistic endeavor. However, it is consistent with the letter and spirit of the amendment that the State can affirmatively support media functions. It already does, albeit in often subtle ways, and has throughout the US' history. Baker observes that some of the favors that US media has long enjoyed from the State include: postal subsidies that better enable media wares to circulate, special access to documents and proceedings, shield laws to protect journalists from legal jeopardy, limited exceptions from antitrust laws (e.g., via Joint Operating Agreements) and from some taxes, in addition to a stream of advertising revenue from the government (1994: xi). These and other boosts to media are established practice and have passed constitutional muster (Baker 1994: 124-62) even if, strictly speaking, they represent some form of State "intrusion" into the sphere of media.

Benson presents further correctives to neoliberal fundamentalism via empirical evidence. To wit, France's openly practiced system of government subsidy for private media has enabled more critical functions than the US system that largely eschews direct subsidy. Benson collected data over several years on an important topic (immigration), with a sample from a cross-section of 15 newspapers (seven French, eight North American). He concludes that "the French press was more critical" than their US counterparts (Benson 2011: 315)—and yet subsidies account for an average of 13 percent of newspaper revenue in France. Moreover, "extra subsidized" newspapers, with greater than usual subsidy, voiced as much criticism as the other French newspapers. The French press was also more multiperspectival than the US papers as it developed a wider range of political positions on the news they report (2011: 315). In turn, Benson speculates that the wider ensemble of French news invites the audience to "think about politics in 'more complex and original ways'" (2011: 315). These outcomes defy simplistic neoliberal dogmas and are clearly favorable to Enlightenment rationality and its brainchild of popular self-rule.

Conclusion

As it stands, US media is heavily privatized and formally independent of the State. At the same time, it has been recruited into

Statist projects when these involve their most drastic, dubious and destructive interventions abroad. In this respect, US media's hallowed independence from and self-ascribed courage with regard to the State is quite often more of a rhetorical posture than a substantive practice. The US' private media system may be construed as erratic watchdogs on, and even enablers of the State, while functioning as "guard dogs barking at the poor" within a pervasively classist culture (Curran 2011: 46).

Subsidy is not the only structural feature that differentiates the US and France. As is the case in Europe more broadly, the French press openly cognizes that it is a participant and player in the political process (for better or worse). In the US, by contrast, the press largely adopts the posture of being chaste political virgins untouched by the corruptions of worldly authority. The seductions of State influence are refused with respect to explicit, no-strings-attached funding. Seductions from the State are, nevertheless, felt through the more covert transaction of what Oscar Gandy (1982) calls "information subsidy." Sophisticated management strategies have also cultivated news media dependence on the power centers of the State and the corporation as significant founts for the sourcing of news; a topic to be developed in the next chapter under the rubric of Herman & Chomsky's third filter.

Notes

[1] See Schudson (2011: 28, 104) for the standard line on MSNBC and Fox News as being symmetrical instances of political slant to the left and right respectively. At present, MSNBC employs an unapologetic woman of the left in Rachel Maddow. It also employs former Republican congressman Joe Scarborough who co-hosts *Morning Joe* from 6 to 9 on weekdays. On his MSNBC webpage Scarborough is described as the author of "*The Last Best Hope: Restoring Conservatism and America's Promise*, a book that draws on the forgotten genius of conservatism to offer a road map for the movement and the country (...) Scarborough inspires conservatives to reclaim their heritage by drawing upon the strength of the movement's rich history" (MSNBC 2011: 1). In other words, MSNBC's programming decisions display ideological balance that has yet to be observed on Fox, notwithstanding ritual assertions of supposed "go for broke" ideological symmetry between the two broadcasters.

[2] One of several comments that follows the Downie & Schudson report exhibits the fevers of doctrinaire neoliberalism: "If we go by Downie's insane and unconstitutional plan to takeover and fund the media, we will get a piece of leftist tripe like the BBC" (Gainor 2009: 33). The comment is standard-issue neoliberalism, entirely orthodox in being colonized, even within the commenter's imagination, by thorough-going preoccupation with the State (and equally through-going embrace of wild assertion over evidence). In this instance, Gainor's *Invasion of the Body Snatcher*-styled claims of Statist intrusion are answered by his call for—cue drum roll—the State apparatus of courts and litigation to police the matter.

CHAPTER 3
Ventriloquism and Other Routines

I want to fulfil my role as a decent human member of the community and a decent and patriotic American. And therefore, I am willing to give the government, the president and the military the benefit of the doubt here in the beginning. I'm going to fulfil my role as a journalist, and that is ask the questions, when necessary ask the tough questions. But I have no excuse for, particularly when there is a national crisis such as this saying—you know, the president says do your job, whatever you are and whomever you are, Mr. and Mrs. America. I'm going to do my job as a journalist, but at the same time I will give them the benefit of the doubt in this kind of crisis, emergency situation. Not because I am concerned about any backlash. I'm not. But because I want to be a patriotic American without apology

—Dan Rather, Journalist (quoted in Jensen 2005: 122-123)

Consider a vignette and what it says about the news environment in one nation. While involved in multiple theaters of conflict, this nation's military ministry recruited dozens of its retired officers and briefed them extensively on current war efforts. The ministry then steered them to television news programs to deliver the talking points under the banner of ostensibly detached expert opinion. The program was characterized by its architects as an exercise in "'information dominance'" through covert-war style "'psyops on steroids'" directed at its own domestic population (Barstow 2008: 4, 7).

Does this program betray the aroma of a vintage Soviet scenario? Perhaps it was a brazen Brezhnev-era act of cynicism as the invasion of Afghanistan bogged down during the 1980s? *Au contraire*: The nation that conducted the covert program of news management was the US during the previous decade. Moreover, the television networks that it infiltrated with brigades of retired military officers were not State-operated broadcasters staffed with orthodox political appointees. In another troubling moment for neoliberal dogma's non-negotiable insistence on State-private sector antagonism, the networks that broadcast the Pentagon-prepped spokespeople were private media stations.

David Barstow's story on the analyst program was showcased on page A1 of *The New York Times* on 20 April 2008. Barstow reports that 75 retired officers were recruited for the program although some participated "only briefly or sporadically" (2008: 5). The retired officers' coaching on Pentagon talking points was underscored with exclusive guided visits to Guantánamo and Iraq after which they delivered commentaries in mass media. Briefings for the ex-officers at the Pentagon parlayed the pomp at the center of military authority. Barstow writes,

> participants describe a powerfully seductive environment—the uniformed escorts to Mr. (Defence Secretary) Rumsfeld's private conference room, the best government china laid out, the embossed name cards, the blizzard of PowerPoints, the solicitations of advice and counsel, the appeals to duty and country, the warm thank you notes from the secretary himself. (2008: 7)

As for results of the program, one retired general with psychological operations expertise toured Iraq and "'saw immediately in 2003 that things were going south'" (2008: 9). Upon returning to the US, however, the same retired general soothingly told Fox News viewers, "'You can't believe the progress.'" The Pentagon's internal evaluation of the retired officers program posited that, "'They have become the go-to guys not only on breaking stories, but they influence the views on issues'" (2008: 10).

Barstow also reports pressure exerted on the retired officers to maintain the Pentagon's line as their media appearances were closely monitored by a subcontractor that was paid "hundreds of thousands of dollars" (2008: 11). If the retired officers deviated from talking points as analysts, they were cut out of the program. Other retired officers may have had still more craven motivations than avoiding exclusion. For example, program participants General Barry McCaffrey and General Wayne Downing "had their own consulting firms and sat on the boards of major military contractors" while drumming up support for militarism (2008: 16).

This news management vignette illustrates the impact of sources since they furnish the building blocks of a news story. Moreover, the contemporary media and public relations (PR) saturated environment furnishes incentive to cover one's tracks in getting spokespeople into the news. The retired officers were "message force multipliers" since they effectively functioned covertly and not as briefed employees of the Department of Defence. They could more easily sway an otherwise sceptical public when presented to the audience as commenting in the dispassionate idiom of former football coaches analyzing a game in which they had no further interest.

The New York Times devoted a hefty 7600 words to Barstow's time-consuming investigation. Barstow was, moreover, commended with a prestigious Pulitzer Prize for Investigative Reporting. Perhaps the system sputtered, one may conclude, but in the end it worked; or perhaps not. While Barstow was garlanded with professional recognition, the story remained largely uncovered and developed no legs under it (Greenwald 2008). Moreover, the *Times'* publication of Barstow's story in April 2008 misses the mark by five or six years. The story was pursued long after it could have usefully informed the public's posture toward the Bush administration's foreign policy template.

Sourcing is the third filter that Herman & Chomsky introduce in the Propaganda Model. As will be explained in further detail, sources provide the diet of information that sustains news narratives; and, in turn, sources are overwhelmingly drawn from the elite sectors of society. However, in elaborating sourcing patterns beyond Herman & Chomsky's account, this chapter will also expand the topic out and devote detailed attention to the journalists' milieu (or "habitus") and their methods. This move follows from the conviction that the material conditions in which journalists work leave residual traces in the resultant news narratives. Whether it is through the everyday seductions of the go-to sources most readily available in the reporter's rolodex—or in being embedded within the military that the journalist is simultaneously covering—news narratives are shaped by "structuring structures" (Benson & Neveu 2005: 3) that are grounded in material conditions.

In making the turn toward analyzing sources and reportorial methods, the question of bias arises—but not in the manner that is more typical of discussions of journalism's performance. In the introduction to a 1994 book, a noted news scholar sighs that he has "grown tired of the enterprise of demonstrating that the media are neither neutral nor are they most of the time 'watchdogs' acting to check the power (...) of dominant interests" (Hallin 1994: 12). The balance of Hallin's book often performs the work of which he is avowedly tired. This chapter will be concerned with making observations about non-neutrality and passive watchdogs, although I take this task as central to an elaboration of the Propaganda Model and not a tiresome matter. However, in offering revision of the Propaganda Model, I will assay a comprehensive account of journalistic methods and practices that in combination constitutes a filter on news content that drives bias deep underground. I will consider the impact of sourcing and information subsidy, implications of objectivity and material conditions ("habitus"), as well as government management of news media. A couple of brief case studies—of the Reagan administration and of embedded reporting from Iraq—will illustrate the concepts. Finally, I address the chimerical implications of new media. Through it all, I assume as Herman & Chomsky do, that journalists are diligent and behave in good faith, but that the "structuring structures" of their work environment inevitably draw their labor into support of the status quo.

Information Traffickers

Sources furnish the constituent elements of stories such as facts, quotes, background, and context. They provide the informational diet that nourishes the story and shapes the contours that it assumes. In his ethnographic account, Herbert J. Gans dwells on the "symbiotic relationship of mutual obligations" and "co-optation" that arise between journalists and their go-to sources (2004: 133, 135). In this view, reporters obtain assistance in crafting stories before a deadline from sources. In turn, sources insinuate their vision of the world into those same stories via facts, quotations, and back story. When relationships develop between journalists and go-to sources, they may also bear the significant, if unanticipated, effect of *inhibiting* at least some forms of reporting. A journalist on a particular beat will be hesitant to report negatively on a valued source to avoid jeopardizing favor in the future. It is hypothesized that, for this reason, the Watergate story was not broken and pursued by White House beat reporters. The story got its legs from young journalists at the city desk who did not depend on maintaining relations with sources within the White House (Gans 2004: 144).

More specifically, who are these sources? With the perspective of a sociologist, Gans illustrates the pre-standing asymmetries in power and social class that impact on sourcing decisions. He writes, "The president of the United States has instantaneous access to all news media whenever he wants it; the powerless must resort to civil disturbances to obtain it" (2004: 119). In other words, established authority always has access to the microphone. In covering economic issues, reporters characterize themselves as substantially more likely to consult figures who speak for government and business than sources from labor unions and consumer advocacy organizations (Goss 2001: 11). A content analysis of sources in *The New York Times*' narratives on an economic issue (North American Free Trade Agreement) exhibits the same pattern that privileges government, business and "expert" economists in constructing the news discourse (Goss 2001). In the light of such findings, it stands to reason that the elites, and their lieutenants, leave a far bigger footprint in the news than advocates for workers or consumers.

Subsidy City

Fashion a press release and the world yawns at the predictable puffery. Garnish the news narrative with quotes and facts in a seemingly independent manner and people may take notice of the same information. For this reason, government, business, NGOs and other actors seek to insinuate their way into the news narrative subtly but palpably. The retired military officers program, considered earlier, is a textbook case.

Techniques by which to insinuate preferred facts into the news were pervasive enough by the 1980s that Oscar Gandy (1982) gave them the catch-all name of "information subsidy." Gandy defines it as "the price of information [that] may be reduced selectively by interested parties in order to increase the consumption of preferred information" (1982: 30). The most straightforward route to subsidizing information is to make it readily available, particularly to time-stressed reporters. The resultant reports are "garbed in a cloak of objectivity" (Gandy 1982: 198) that makes the subsidized information more potent for having been stitched seamlessly into the news. The methods by which to subsidize reporters' time and efforts are legion. They include on-the-record interviews, press conferences, off-the-record backgrounders, news releases, and reports with executive summaries that cut to the main claims. The private sector generates information subsidy through Video News Releases (VNRs) designed for broadcast on television news programs. A VNR that originated with military contractor Lockheed Martin on its F-22 fighter jets was seen by an estimated 41 million US viewers when it was re-packaged into a news story (Bennett 2001: 9). Release of a department store's Christmas catalogue (91 million viewers) also transformed spectacle and "pseudo event" (Boorstin 1992) into a story broadcast under the rubric of news. Since Gandy indentified the concept of information subsidy, think tanks have proliferated with a mission to incite discourse and channel information that they have produced into public discourse that includes the news hole (Goss 2006; Soley 1995).

Reagan-era Deputy Press Secretary Leslie Janka explains that, "'As long as you come in there every day, hand them [journalists] a well-packaged, premasticated story in the format they want, they'll go away.'" Janka adds the punch line: "And you do that long enough, they're going to stop bringing their own stories and stop being investigative reporters" (quoted in Hertsgaard 1989: 52). Practical considerations also undermine opportunities for journalistic investigation. The work "required for an exposé is expensive and not always productive, for reporters must usually be assigned to the story for weeks, if not months, thus making them unavailable for other stories" (Gans 2004: 118). Gans adds that "sometimes, months of investigation may not produce a suitable story." These opportunity costs give further advantage to information subsidizers' efforts to reach reporters.

Objectivity and Its Implications

The recipe looks straightforward. Under the doctrine of objectivity, one consults a variety of sources and thusly guarantees the construction of a balanced story. Notwithstanding constant right-wing fist-shaking about "crypto Marxism" and slippery slopes in every headline, the method of objectivity "works" in the lights of rigorous studies of it. After sifting through data, Entman answers right-wing histrionics about the news by positing that, "there is no acceptable intellectual basis for concluding that the national media consistently promote the left or indeed any view" (1989: 36).

At the same time, "important biases in the news occur not when journalists abandon their professional standards but when they cling most responsibly to them" via the "establishment slant" of prevailing norms (Bennett 2001: 182). Moreover, the routines of objectivity and its prescription of "two sides to every story" may "actually homogenize the political content of reporting" (2001: 151). A story may indeed have fewer or more than two viable sides! However, assuming two sides has the commercial advantage of assuring a degree of narrative tension in every story (he said/she said), even if it is contrived. Objective reporting also re-circulates "familiar slogans and ideological precepts" associated with the two sides and thus privileges them once again (2001: 184). By contrast, European press is less beholden to a two-sided objectivity model and regularly exhibits a more open, uninhibited press culture.

Three Spheres

There is evident tension between consensus and dissent within the news. In accounting for this in the light of the claims made above, imagine three spheres. In the first, the "Sphere of Consensus," matters have been widely agreed. Within it, the methods of balance and objectivity are largely mooted as all players are understood to be on the same side of an issue. "Motherhood and apple pie" reside here as uncontroversial objects of universal acclaim (Hallin 1994: 53). Patriotism, individualism, meritocracy and capitalism are also nourished within the sphere of consensus since their premises are given, largely without demands for argument or evidence. Legitimation is reflexive!

Although the boundary may be amorphous, the sphere of legitimate controversy abuts the sphere of consensus. A high proportion of news plays out in this sphere. Hallin (1994: 53-55) cites election coverage as a paradigm of legitimate controversy since campaign stories are made to order for objective journalism and exquisitely balanced claims and counter claims. Moreover, candidates from the two main US parties often agree on broad assumptions (consensus), such as the primacy of the US in world affairs. However, they often diverge when considering the techniques by which to realize agreed objectives. These legitimate controversies may revolve around the calibration of aggression versus "soft power" within the deeply held consensus that the US must be preeminent and interventionist across the globe.

Pluralistic political systems all but guarantee that legitimate controversy will be an everyday occurrence. In this view, competing parties will loudly voice distinctions from each other even as they are circumspect on their degree of consensus. However, Hayes & Guardino (2010) argue that the ostensible opposition party, the Democrats, largely kept their heads down and their criticisms muffled in the run-up to the US invasion of Iraq in 2003. They claim that US journalists therefore sought out dissenting international opinions to generate a modicum of controversy, even though international sources are generally less trusted by US audiences.

Finally, beyond consensus and controversy, there is a sphere of deviance. As in the sphere of consensus, argument is not needed to deal with the deviant. Recognized extremist ideologies (Nazi, communist) get no traction in mainstream media. Hallin points out that, while the boundaries between the three spheres may be vigorously patrolled by journalists, social movements can and do reconfigure those boundaries. Hallin cites the US' 1960s anti-war movement as an instance of a discourse that exited the sphere of deviance to become ensconced within legitimate controversy. One can argue that economic neoliberalism achieved something similar during the 1970s (Harvey 2005: 39-63; Klein 2007).

Journalism Under the Golden Arches

Reporters require sources and resultant news discourses tend strongly to resonate within the spheres of consensus and legitimate controversy. As this plays out, what institutional practices and constraints govern journalists' work? In detailing an answer, I will rely on scholarship from the UK that has focused on the problem.

In an important monograph, Justin Lewis and colleagues at University of Cardiff examined sourcing patterns within "hard news" at the quality end of the British press continuum. Their findings suggest that even the British broadsheets are strikingly dependent on material from informationally subsidized sources external to staff's journalistic efforts (Lewis, Williams, Franklin, Thomas & Mosdell 2006). Specifically, 50 percent of the broadsheet papers' news stories were "almost directly" derived from "or largely dependent" on wire service content (2006: 15). Although "public relations activity—particularly the more sophisticated kind—may leave few traces," another 20 percent of news content was generated out of PR materials according to the researchers' investigation (2006: 13, 17). Lewis, Williams, Franklin, Thomas & Mosdell estimate that only a thin slice of newspaper content—to wit, 12 percent—is a result of journalists' "nose for news" and developed via independent investigative initiatives. In making their case, they present several examples in which they trace content in its voyage from the press release to the printed page in British newspapers (2006: 29-41). As a result, a significant share of partial, soft-focus puffery is regularly insinuated into the news ecosystem. These trends also re-circulate and reinforce the authority of already powerful actors (from the State and private sector) as they possess more resources to employ the PR means of discursive production. Given that journalists often make minimal transformations of prepackaged source materials, the researchers also characterize the widespread practice of taking up massive information subsidies as "plagiarism by another name" (2006: 28).

At the risk of seeming nostalgic about past journalist practices and "golden ages" (that may have been more like a pale shade of yellow), renowned investigative journalist Nick Davies concludes that the Cardiff investigators furnish evidence that "something fundamental has changed" in journalism (2009: 53). In addressing what has changed, Davies has coined the term "churnalism" to account for modestly paid reporters manufacturing high volumes of articles to fill a growing news hole. He cites the case of a reporter who, by the diary, composed 48 stories during a normal week, or more than one per hour. At the same time, the reporter spent only 3 hours outside the office and met 4 people face to face across the same week (Davies 2009: 56-59). Under the regime of churnalism, it is wired into the system that reporters will depend upon taking information from where it can be most easily harvested, such as the internet, press releases, and communication officers for the ministry or company.

Surveying similar trends, Bob Franklin (2005) has minted the term "McJournalism" via analogy between the de-skilling of food preparation and trends in reporting. In particular, the fast food industry is organized around "the principles of efficiency, calculability, predictability and control" (2005: 137-138). Adherence to these top-down prerogatives account for fast food restaurants' attainment of uniformity and, finally, profitability. Where calculability is concerned, reporters at the US-based Gannett Group work under some very particular prescriptions. For one class of story, reporters are instructed that they "'should use one press release and/or one or two cooperative sources, should spend 0.9 hours [54 minutes] to produce each story and should deliver 40 such stories each week'" (quoted in Franklin 2005: 145). Would the reader like that article with fries to go? Under the "'pre-formed, pre-cut, pre-sliced'" regime of fast food-styled news, de-skilled workers microwave prefabricated product such as press releases and serve them into the news hole.[1]

Metro, a free tabloid and sixth largest paper in Britain as of 2005, employed six reporters in London and four in Manchester to produce its product (Franklin 2005: 148). However, speed-up is evident in even the UK's central and prestigious news organs. Although it is not a soporific epoch for news, the British Broadcasting Corporation (BBC) has made several rounds of steep reductions in staffing (Davies 2009: 67). An internal BBC memo from 2005 calls for a four-paragraph story to be written, fact-checked and forwarded to the news desk *within five minutes of the news breaking*! The memo also calls for ten paragraphs on breaking news to be composed within 15 minutes. Stories may or may not get out on time to these specifications. However, such policies have succeeded in generating a legion of BBC stories that have not been fact-checked (60 in one day, according to an internal study) and that are laden with the grammar errors, misspelled names and botched details one expects of work done in haste. Errors of commission aside, it is not possible to jig-saw together the "big picture" of the news while under such time constraints.

Angela Phillips (2010) complicates the picture by noting that more experienced and higher prestige journalists are partly exempted from the general speed up. Established journalists, "with the highest level of autonomy" from management and market-induced time pressures, "regularly have the luxury of phoning a number of different people to verify information, or probing for alternative views" (Phillips 2010: 100). This finding betrays a curious notion of what constitutes reportorial "luxury" in contemporary journalism. Nonetheless, many (particularly younger) journalists are "asked to produce up to a dozen stories a day." It is not surprising that they have difficulty thinking of "a single recent incident of a self generated story" when they are regularly pulling stories off the wire and from other subsidized sources (Phillips 2010: 91).[2]

The Journalist's World

In addressing the practices of contemporary journalism and journalists, one also confronts the romanticized idea that reporters are oppositional figures with shrines to Nelly Bly, I.F. Stone, and Anna Politkovskaya in their document-cluttered apartments. In his classic ethnography, Gans registers his doubts about reporters as iconoclasts. He observes that journalists,

> have more power than the rest of us, but mainly because they express, and often subscribe to, the economic, political, and social ideas and values which are dominant in America. Indeed, as much as I was writing about the journalists, I felt my book was as much about the dominant culture in America, and about its economic and political underpinnings, as about them. (2004: xxv)

Gans' subsequent content analysis backs these claims since presumptively universal middle-class values (sphere of consensus) drive the news.

In contrast with the image of news media as permanent opposition, consider 1980s ABC *World News Tonight* producer Jeff Gralnick's characterization of his work: "'It is not my position to say, "For shame, Public Agency." It is my job *to take the news as they choose to give it to us* and then, in the amount of times that's available, put it into the context of the day or that particular story'" (emphasis added, quoted in Hertsgaard 1989: 62). Gralnick concludes that, "'The evening newscast is not supposed to be the watchdog on the government'" and adds a dollop of teleological certainty that, "'It never was, never will be'" (1989: 62). It stands to reason that a news organ will not conduct investigations or report on government wrongdoing without the volition to do so. Abu Ghraib prison abuse (Allan 2005: 1-4) and the Bush administration's warrantless wire-tapping program testify to journalism's passivity in the face of pressure to suppress reporting on even criminal State-backed activity. As concerns warrantless wiretaps, *The New York Times* cooperated with the Bush administration in sitting on the story for a year (CNN.com 2005) although it would later publish editorials denouncing the Bushites' practices that it had previously condoned through silence when pertinent laws suits gained traction (*New York Times* 2010).

Why is news media frequently complacent about and submissive to State and corporate power? Drawing on Bourdieu's concept of the "'structuring structure'" of habitus, Benson & Neveu argue that "the individual's predispositions, assumptions, judgments and behaviours are the result of a long term socialization" and include "professional education" (2005: 3). One might quip that habitus is matter of everyday habit, acted out in practice in the material milieu. In this respect, Bourdieu's conceptual apparatus resembles influential theories of ideology in its orientation toward hands-on action (Althusser 1994; Eagleton 1991). In practice, "structuring structure" means immersion in an environment laden with subtle conditioning, particularly for the Washington-based reporter. The seductions of "the perks, the trips, the *life*, as they say, are more than any human being should be expected to withstand" without becoming a creature of it (original emphasis, Washington correspondent Thomas Oliphant, quoted in Hertsgaard 1989: 59-60). In their work, journalists "engage with first and foremost those agents who possess high volumes of (social, symbolic, economic) capital" (Benson & Nevreu 2005: 5). For established journalists, this tendency toward reliance on the powerful may be the result of conviction. For younger journalists, such dependence may be a matter of convenience. In either event, the outcomes are similar in valorizing establishment sources.

Journalism is implicated in the cosmology of power. It is, however, a subordinated part of that cosmology, like a planet that reflects its lodestar and does not generate its own light. As a consequence of journalism's subordinate role, paradoxically, there are some freedoms for "deviant trajectories" that journalists may take up (Benson & Nevreu 2005: 6). In this view, journalists may go "deviant" more easily than heads of major banks who are more centrally placed (and constrained) within existing power relations. Such deviant reportorial trajectories include the careers of notables such as Barstow. In the main, however, journalism is proximal to ruling assumptions and practices. In a neo-liberal dominated environment, it also follows that news media is pulled "closer to the commercial pole in the larger field of power" (Benson & Nevreu 2005: 6).

The *Posts*: A Tale of Two Cities

Journalist William Greider has pursued a deviant reportorial trajectory as an outspoken advocate for the working class. His reflections on his career illustrate the impact of reporters' materially grounded habitus. Greider compares two newspapers, both called *The Post*, at which he has worked. The first *Post* is located in Cincinnati and the second in District of Columbia. At the *Cincinnati Post* in mid-twentieth century, the city room "resembled an industrial space more than an office" with "pneumatic tubes and piping (...) exposed overhead" (1992: 288). The *mise-en-scene* of the *Cincinnati Post*'s office was revealing since the reporters were not college educated but had grown up in the "Queen City" and knew it intimately. They also knew all their colleagues as "there was no social distance between the newsroom employees and the *Post's* printers and pressmen—they were all working class" (1992: 289). Greider acknowledges that the resultant Cincinnati paper had shortcomings. Nonetheless, in contrast with Murdochian simulations of populism fronted by bizarre millionaire celebrities such as Glenn Beck and Sarah Palin, the *Post* "was frankly and relentlessly 'of the people' and it practiced a journalism of honest indignation on behalf of their political grievances" (1992: 290).

Later, Greider worked at the more prestigious *Post* in Washington, the reputation of which has long been fortified by its pursuit of the Watergate story. The difference between the two *Post*s speaks to both intensifying professionalization in journalism and the turn toward servicing the more comfortable class demographic. In Greider's words, "A trade [reporting] that had once been easily accessible to the talented people who lacked social status or education was converted into a profession" (1992: 289) with the *Washington Post* at its apex. Its reporters are tolerant, cosmopolitan, polished—and "more comfortable dealing with people in authority, given their backgrounds, but not necessarily more critical" than the Cincinnatians from whom Greider learned the trade (1992: 294). In media center cities, journalists do not know the working class support staff in their own building, but "justifiably regard themselves as social peers of the powerful figures whom they cover." The resultant journalism is, in Greider's withering appraisal, "Strong on facts and weak on truth" (1992: 306). Moreover, by 1992, Greider characterizes the contemporaneous *Cincinnati Post* as having succumbed to being a chain newspaper and attuned to an upscale demographic. It is one of many city papers that had become "bland shadows of their former selves" and that tolerate fewer deviant trajectories (1992: 292).

Reaganite Media Management: A Case Study

Thus far, I have surveyed issues that come to bear on the contemporary shape of journalism (sources and subsidy, objectivity, churnalism, habitus) and that relate to reportorial routines. Now, I will try to pull them together via a case study of the Reagan administration before orienting to more recent developments in news management.

While news is riddled with "contrived and managed news performances of entrepreneurial politicians" (Bennett 2001: 27), Reagan's administration perhaps epitomized these tendencies. It also exhibited a facility for news management that far outdistanced the Carter administration that preceded it (Entman 1989: 53-62). In light of this facility, the administration and its figurehead were able to fashion assertions about reality that often bore no resemblance to it (Chomsky 1988). In its entrepreneurial mode, the Reagan team sold an image of its own governing efficacy to both US parties and the nation's public. As a result, Reaganite ideology (neo-liberal, allegedly "small government" Statism) has occasioned continuing and deleterious impact long after its term of office.

Mark Hertsgaard's account demonstrates that the Reagan team's media management was effective in exploiting the seams within journalists' methods and routines. With sophisticated comprehension of press behaviours and relentless discipline, the Reaganites managed journalists as if they were "'a bunch of caged hamsters thoroughly dependent on their masters for their daily feeding'" (reporter Walter Robinson, quoted in Hertsgaard 1989: 52). In less colorful terms, Hertsgaard posits that Reagan's team was particularly effective at "using the press" as "an unwitting mouthpiece of the government" (1989: 5).

Pillars of Reaganite media management included, "Talk about the issues *you* want to talk about," "Speak in one voice," and "Repeat the same message many times" (original emphasis, Hertsgaard 1989: 34). The practice of developing and disseminating the line of the day pulled all three imperatives together. The line—"inflation is falling!" "the Sandinistas are menacing America!"—was set by 8:15 in the morning by a core group of officials that included James Baker, David Gergen, Richard Darman and Larry Speakes. The line that they selected each day answered to their over-arching quest—"'What are we going to do today to enhance the image of the President?'"—by cuing the news media's attention (1989: 35). The line was then transmitted to other senior officials and to spokespeople throughout the federal apparatus. By 9:15 each morning, reporters had the line in hand while Reagan's team assayed to push it up into the evening news programs to make them, in effect, shop windows for the administration's priorities.

The central news management technique was to recruit reporters to "spread the word" for the administration of their own volition in the course of doing their jobs. Thus, the Reaganites did not suppress journalists, but co-opted them while coaxing their discourses into the administration's preferred frames of reference.

Polling was instrumental to the strategy. Lead pollster Richard Wirthlin determined which policy positions where resisted by voters and what messages would blast through recalcitrance. If and when the public had come around, polls also furnished this information. To take a concrete example, Reagan's cuts to federal aid to education were initially opposed by the public by a stark 2 to 1 margin. Rather than change the policy, an offensive was ordered by Reagan's core group of communication advisers. The figure was eventually inverted to *2 to 1 support* for the cuts. A third of the electorate was moved by the issue campaign that featured massive repetition when Reagan delivered his lines, but without policy concessions (Hertsgaard 1989: 48-49).

The Reagan White House's reliance on "photo ops" lent further support to the cuts campaign. In one network news report, Reagan was characterized as unable to answer a basic question about the education policies that he had traversed the country promoting, as he deferred to his Secretary of Education for an answer. Reagan's own team was unfazed since, in the same report, impressions were tinted by the president in a classroom surrounded by children. Reagan's team held that careful staging of visuals could countervail even critical commentary. By contrast, Reagan's Democratic opposition was relatively indifferent to visuals and journalists' daily need to file stories. According to a CBS News producer "'Republicans would talk to you about logistics from the start because they wanted the picture out as much as you do'" (quoted in Hertsgaard 1989: 42).

Attenuated flak through pressure on news workers was also in the toolkit. Gergen, the administration's Director of Communications, had the practice of calling each of the major networks in the hours before the evening news to discuss upcoming stories. Gergen made no censorious gestures or threats that would have backfired given US news media's strong legal protections (and its aggressively maintained, rhetorical postures of independence). However,

CBS Evening News producer Tom Bettag characterized the calls from Gergen as "no small thing" even if they dickered "usually over relatively small details" (quoted in Hertsgaard 1989: 31). By their own accounts, it is likely that the news workers subtlety tailored the evening news programs in anticipation of what they would have to say to Gergen.

Along with the stick of carefully apportioned flak, Reagan's team worked hard to flatter the press. During Reagan's first three years in office "more than 150 special White House briefings were organized" (1989: 37). During these briefings, 80 or 90 news workers—journalists, editors, presenters, news directors—from across the regions of the US were invited. In turn, news workers from peripheral parts of the country "'were thrilled to come'" to the nerve center of governance in Washington (White House official, quoted in Hertsgaard 1989: 37). Once there, the news workers were greeted with a program that featured a cabinet officer or an undersecretary, followed by a lunch attended by Reagan in the state dining room. Regardless of what was served as a libation, the effect was intoxicating and registered on the press before they returned to work to file or edit stories. The kindly treatment also presented a stark contrast with the far less indulgent approaches taken by Reagan's predecessors Nixon and Carter. Alongside the attention to news workers from the periphery, Reagan and his team were friendly and solicitous with reporters in the Georgetown party circuit. Once again, the tendency contrasted with the more introverted postures of Nixon and Carter and their teams (Hertsgaard 1989: 44-45).

In such ways, the deck was stacked by Reagan's team in favor of getting a thorough hearing and the benefit of the doubt from the media. In their role as professional mediators, journalists constructed a favorable prism through which the government was regarded by the public. Reagan was not forced into the desperate tactic of "talking over the heads" of the media by making prime-time television speeches to the public, discourses that occur during a thin slice of time during while the public may not be listening. On a day-by-day basis, the media was subtly co-opted into the project of making Reagan and his broader project appear to "stand tall" on the political landscape.

Beyond Reagan

Surely, one may counter, much has changed since the Reagan era. Due to post-Cold War conditions and the impact of new technical substrates, we must be living within an increasingly unscripted current of events. A wizened press corps "Won't Get Fooled Again" (Townsend 1971) and cannot be herded any more than cats can. The Clinton era suggests some highly qualified support for this thesis. Following a tendentious relationship during the 1992 general election campaign, Clinton's team entered the White House wary of news media. The administration alienated news workers further by closing the hallway between the press room and the White House press office while attempting to go over the media's head via tactics like "electronic town halls." A tense situation prevailed until the more emollient Mike McCurry's appointment as Press Secretary (Bennett 2001).

Despite the warps in its news management, Clinton's team nonetheless took leafs from the Reagan team's manual to regiment the march of events. Specifically, the Reaganites strategically toured the president through the nation's heartland to gain coverage by local media that was awed by a sitting president's visit. The locals often agreed not to ask proscribed questions and to carve interviews into five-day tranches that, in turn, generated a full week of favorable news discourse (Hertsgaard 1989: 49-51).

Similarly, in the days immediately after Monica S. Lewinsky had become a household name in January 1998, Clinton visited Urbana-Champaign, Illinois, with an entourage that included Vice President Al Gore and Secretary of Education Richard Riley. As central Illinois' "twin cities" are two and a half hours from Chicago by car, with a population of approximately 130,000, one might ask, "Why Urbana-Champaign?" The media management technique in this case moved the political field of play into the nation's heartland. In doing so, it garnered more supportive coverage in local media than the frenzied Washington-based media was channeling. Urbana-Champaign is also unusually well situated for the purpose of reaching multiple laminas of local media. In the Assembly Hall parking lot that morning (28 January 1998), I observed a fleet of news trucks from Chicago, Saint Louis, and Indianapolis (all within 3 hours by car) covering the "local" visit of the sitting president.

Even *The* (Champaign County) *News Gazette* suspended its habitual anti-Clinton discourse in deference to the "flattering" visit.

From Teflon to Kevlar: In Bed with Embeds

More recent evidence demonstrates that journalism did not become more sceptical following the Reagan experience. After Clinton, the George W. Bush administration occasioned startling degrees of obsequious news coverage, particularly during its first term. Rather's epigram to this chapter serves as one bombastic example. A careful and extensive study by Hayes & Guardino (2010) further testifies to the extent to which news media deferred to the Bushites.[3] Hayes & Guardino's content analysis of 1,434 stories from evening news programs indicates that the Bush administration furnished 28 percent of all quotations broadcast in the run-up to the invasion of Iraq. Bush alone, in "tough guy" warrior mode, furnished 15 percent of all quotations (2010: 72). By contrast, and despite "what were likely the largest pre-war protest actions in world history," anti-war groups generated 1 percent of all quotes (2010: 74). Forty percent of quotes critical of the incipient invasion that were broadcast by the networks were attributed to Saddam Hussein and Iraqi officials. Their credibility with the US audience was nil, but they were nevertheless pressed into service as the news narratives' leading doves. Hayes & Guardino acknowledge that, although their methods were "constructed to capture even faint signals of dissent," such as circumspect tactical disagreements, news discourse on administration aggression toward non-belligerent Iraq was clearly positive (2010: 66).

In the aftermath of the 1991 Gulf War campaign against Iraq, Hallin comments that, "the power of the state to manipulate public opinion was mediated through populism" in which "the lives of the soldiers were put on the line." Following the populist turn, "television and the public at large identified with them [soldiers] and their families, critical distance collapsed, and media and public opinion were integrated into the war effort" (1994: 14). Hallin's characterization of Gulf War media populism contrasts starkly with a sample during the US' aggression in Vietnam in which 63 percent of quoted military sources were officers and only 13 percent were enlisted soldiers (Gans 2004: 120).

During the 2003 invasion, populism was put in motion through embedding as "more than 600 journalists working for news agencies from around the world traveled alongside U.S. and coalition forces as they invaded Iraq" (Linder 2008: 1). In a book-length examination, Justin Lewis, Rod Brookes, Nick Mosdell, & Terry Threadgold (2006) find that the embedding program was driven by PR-friendly motives to stimulate a steady stream of invasion-affirmative news into the prevailing 24 hour media cycle. That is, US troops were in effect cast in a "militainment" drama to satisfy the strategic objective that "the face of the war should be America's sons and daughters" in uniform and on patrol in Mesopotamia (Lewis, Brookes, Mosdell, & Threadgold 2006: 54, 51). Lewis and his colleagues are careful to point out that embedded journalists did in fact submit reports that were critical of the military and made observations not readily available unless in close proximity to the forces. An example on which Lewis and colleagues dwell is a fatal shooting of seven Iraqi civilians at a US-operated checkpoint, authoritatively reported by an embedded journalist (2006: 140-143). However, the upshot of their multi-method investigation is that the embed program functioned as planned by the Pentagon and often cast the soldiers as the news' protagonists. In doing so, news reports hewed closely to the military's favored narratives. Even reports of the fatal shooting at the checkpoint align with the PR goal of adopting the posture of transparency and owning up to mistakes, in exchange for generally getting one's narratives across to the audience through journalistic intermediaries.

In Lewis, Brookes, Mosdell, & Threadgold's critical appraisal, embedded reporters were not obviously censored. Nonetheless, the material fact of embedding fostered deep continuity of interest and complicity via "dependence of correspondents on their units for food and water, transport and safety" (2006: 94). Following their converging methods, Lewis, Brookes, Mosdell, & Threadgold conclude that, "the military and government understand very well how the media works and have moved beyond this 'censorship' model (...) to embrace a public relations model. This has less to do with *preventing negative* coverage than *creating positive* coverage" (original emphasis, 2006: 195). They add that the military/government seduction of media workers gained traction "not because of any *failure* of normal media practices, but precisely because *professional journalists were carrying on with business as usual*" in being recruited into a public relations-style of reporting that followed the lead of their sources in uniform (original emphasis, 2006: 197).

In a content analysis of embedded reports during the invasion, Andrew M. Lindner (2008) corroborates Lewis, Brookes, Mosdell, & Threadgold's (2006) account. Across 742 news articles, Linder compares the content generated by reporters who were embedded in military divisions with those who were not. While it is not surprising that US reporters are inclined to take the US side—it would be curious to expect otherwise—Lindner discusses the techniques that cemented the journalists' loyalty. Reporters attended a week-long "Embed Boot Camp" where they were informally inducted into the military. Specifically, they were "outfitted with Kevlar helmets and military garb, slept in barracks bunks, and ate military grub in the mess hall" (2008: 4).[4]

Along with embedding, Linder observes that some US news organizations went the more costly route of stationing reporters in Baghdad or gave reporters "the freedom to roam" elsewhere through the Iraqi theater (2008: 5). The contrast between embedded and non-embedded reporting forms the crux of Linder's investigation and dovetails with Lewis, Brookes, Mosdell, & Threadgold's (2006) conclusions. To wit, Linder's content analysis shows that embedded reporters who share life in the trenches with the soldiers will file reports saturated with details drawn from that perspective. These same journalists could, of course, be roused to being oppositional if reporting on the Secretary of Defense's press conferences. However, reports authored at the soldiers' side give the perspective of the Department of Defense's employees on the ground as they implement the Pentagon brass' orders. It is a seductive method of news management that eschews "heavy handed propaganda" (Lindner 2008: 7), since the Pentagon's invasion project is insinuated into the news with powerful if implicit sympathy for being channeled through the front lines. Moreover, this strategy does not suppress reporting. To the contrary, it seeks to incite more and more reportage, given that it is anchored in sources and conditions that readily fold into the Pentagon's narratives.

"While the embedding program didn't only print good news," in 93 percent of embedded reporters' articles, soldiers implementing the Bush/Cheney/Rumsfeld invasion were used as sources. This figure far outdistanced the proportion of articles using soldiers as sources that were composed by the independent and Baghdad-based press (and by 2- and 4-fold). Almost 40 percent of the embedded reporters' stories bathed soldiers in the human interest light. By contrast, the non-embeds composed human interest narratives on US forces in a microscopic proportion of stories (less than 1 percent).

Although one of the rationalizations that the Bush administration touted for the invasion was its alleged benefits to the Iraqi people, the citizenry is far less visible in embedded journalists' reports. Iraqi sources appeared in 22 percent of their stories while human interest on Iraqis' situation constituted 9 percent of the embedded news narratives. The independent and Baghdad-stationed reporters were three to five times more likely to report from a human interest perspective grounded in Iraqi civilians' experience. As to the obvious problems generated by a large scale invasion, such as civilian death and destruction, the non-embedded journalists were two to four times more likely to include such civilian disasters in their reporting than embedded counterparts. Consistent with Lewis and colleagues' investigation, Linder presents clear evidence that news management can shape sourcing patterns through strategically constructed access.

Clockwork Oranges? Or Is This View Too Mechanistic?

Surveying the landscape of news, one could point out that today's newspaper or newscast is cloyed with raging controversies, ideological clashes and tough postures toward public and private sector authority. Of course, something like that is inscribed on newsprint and screens. The years of saturation coverage of Clinton's libido, to take one example, does not suggest a news industry that rolls over for authority figures (Bennett 2001: 22-23). However, one must consider what kind of adverserialism may be at turns docile and viciously biting.

When both reporters and politicians are invested in their roles, they will inevitably clash with each other (Bennett 2001: 192-195).

Just as "good students" may perform the favor of asking questions that invite the professor to disagree and "correct" them, reporters similarly engage with ritualistic acts of opposition. Once the opposition is quelled, order is restored and the legitimacy of authority invigorated. While palpably real for participants and their sense of integrity within respective roles, surface conflict between journalists and sources is strongly tempered by "underlying value consensus" (Bennett 2001: 194). In a related vein, Israeli scholars Neiger, Zanberg & Meyers (2010) develop the concept of "reaffirming criticisms." Such news discourse assumes the central premises of authority while venturing criticisms around the edges to the effect of "Israel is not Israeli (or tough, or self-interested) enough."

Bennett claims that adverserialism "is more often personal than substantive" (2001: 150). For instance, news discourse on investment fraudster Bernard Madoff posits him as *a bad man* who needs to be punished, an appraisal that is quite true in what it explicitly asserts. However, what is left out of the statement? The ecosystem of deregulated casino capitalism in which Madoff operated is under-reported when adverserialism orients to the personal over the substantive and systemic. Bennett holds that there is inherently nothing wrong with news about particular people. Stimulating melodrama and Greek tragedy are about people too! Bennett adds, however, that the distinguished exemplars of drama—and of journalism—execute a pivot from the particular to the general.

Adverserialism that fetishisizes the personal is also continuous with "gotcha" journalism directed at personal flaws of public figures. Clinton's appetitive sex life was, by any sane standard, less consequential than his administration's pursuit of punitive sanctions that eviscerated civilian life in Iraq during the 1990s. Estimates of civilian mortality due to the stubbornly maintained sanctions regime reached over one million while the infrastructural backbone of Iraqi civilian life was decimated and never reconstructed after the Gulf War of 1991 due to the crippling sanctions (Mueller & Mueller 1999; Pilger 2000). Despite the shocking scale of punishment directed at civilians, the sanctions on Iraq were answered with meagre coverage that colored within the lines furnished by Clinton administration narratives (Goss 2002).

Gotcha journalism contrasts with "solid investigative reporting aimed at looking for stories with substance and enduring importance" (Bennett 2001: 158). As it does not demand the time and resources that are needed to dig a story up by its roots, gotcha journalism also enjoys the advantages of being easy and low cost. Throwing lazy spit balls and moralizing epithets from the grandstand will suffice. Moreover, if getting stories into print is a metric of how well a reporter's career is progressing, the incentive for gotcha journalism is reinforced at the expense of being an investigator.

The gotcha tendency does aggravate and threaten particular figures who may, in turn, be readily replaced by a party's roster of "back benchers" should they fall victim to scandal. By contrast, substantial adverserialism is a threat not merely to wayward individuals, but to established institutions and entrenched practices. In these ways, substantial adverserialism presents complications for journalism since, as currently constituted, news media is dependent upon elites and the established mode of operations.

New Media Interventions

The account of prevailing news routines and sourcing patterns presented above posits that they effectively insure systematic warps in apprehending reality. Given this and similar critiques of news, the advent of the internet should be cause for uninhibited celebration. With its avenues for alternative viewpoints and audience participation, is new media not the one-stop solution that has been long awaited? One answer to the question is: Not so fast. Phillips reports from the frontlines of an extensive investigation that "every journalist interviewed was effected to some extent by the need for speed and greater output" in the new media environment (2010: 90). She continues that, "One is forced to conclude that the overall effect of the internet on journalism is to provide a diminishing range of the same old sources albeit in newer bottles" (2010: 101).

Before considering problems heralded by new media technology, some words are in order on its upsides. Stories that previously would have been difficult or impossible to cover are now captured on mobile phone. In this fashion, British reporters discretely covered authoritarian Zimbabwe's 2008 electoral charade (Lee-Wright

2010: 72). "Email is phenomenal," raves another journalist, since "It just speeds things up" and helps one to stay ahead of the curve (quoted in Davis 2010: 130). Email also means that a reporter does not have to arrange meetings to pick up documents while internet convergence enables him or her to watch legislative activity without leaving the desk. Another journalist enthuses that new media's spotlight on "most viewed stories" enables work from home since ideas for articles gravitate to the journalist via the internet connection (Tremlett 2011). While reporters' work may have sped up, this is countervailed by the warehouses of information that have never been as readily available. Moreover, it is important to recognize that "the web also allows unprecedented access to dissenting voices" (Phillips 2010: 98). This addresses a problem that has long been significant for those voices that are beyond the perimeter of the mainstream, as Gitlin has discussed (1980).

Speed Racers

Despite important developments that break from the past, new media has in some respects reinforced the old. As noted in Chapter 2, new media economics are vexed by the "'haphazard and fragmented way'" that readers approach internet news as compared with reading a newspaper hard copy (Freedman 2010: 45). Hence, the internet generates mere "'pennies for dollars'" for its glut of content and readers' fickle grazing patterns suppress advertising revenues (Freedman 2010: 45).

The contemporary journalist has been characterized as being a long way from Clark Kent: "desk-bound (and) internet captive" (Fenton 2010b: 166). While diligent labor is surely performed at the desk, the reporter is in an environment that has "'broken the link between the unique relationship part of journalism, which is number one of what journalism is about'" (newspaper section editor, quoted in Philips 2010 93). A single face-to-face briefing tends to generate several worthwhile stories for seasoned reporters and makes meeting "in real life" more time-effective than internet-enabled news gathering methods in some respects.

At present, the internet contributes to underinvestment in journalistic labor and in the intensification of churnalism. The news hole is bigger by a factor of two or three, depending on whose study is consulted from what timeframe (Davis 2010: 129). Deadlines are almost a quaint anachronism when the drive to match competitors' updates of their web pages is perpetual. It is now common for reporters in the UK, with management's encouragement, to "rewrite stories appearing elsewhere in some cases without a single additional phone call, and to lift quotes and case histories without any attribution" (Phillips 2010: 96). While this rewriting practice is not overtly tolerated in the US, American news aggregators engage in substantially the same practice but eschew the pretence of rewriting.

New media advocates also posit that groups outside the elite circle will be able to penetrate into the news discourse with the advent of the un-filtered internet. While the success of some bloggers provides significant affirmations for these possibilities—for example, the "out of nowhere" emergence of Joshua Marshall, Marko "Daily Kos" Moulitsas, and Glenn Greenwald—Fenton finds a more equivocal record of new media breakthrough for NGOs. While the internet "may provide constant *possibilities* for the fracturing of dominant discourses," in practice NGOs use new media as merely another channel for their messages (original emphasis, 2010b: 162). In order to be noticed and taken seriously, workers in NGOs are trained to fashion messages to specifications that will allow them to pass into the mainstream media. Fenton fingers the paradox: "if they continue to mimic the requirements of the mainstream" with objective sounding messages, "then arguably they will fail in the role of advocacy and become no different to elite sources of information" (2010b: 167). NGOs even face new expenses as a result of the internet. As one NGO employee explained to Fenton, they must "'put a lot of money into (...) improving our website'" since today without a decent looking one "'you're not in the game'" (2010b: 163). In this view, irrespective of the internet's pronounced upsides, a new technological platform is not in itself sufficient to rearrange the pre-standing alignment of power relations—and, in at least some respects, may even reinforce them.

Conclusion

This chapter stakes out and elaborates the commonsense position that sources who furnish reporters with information, back story and quotations significantly shape the form of the journalistic narrative. In this view, there is a frequent pattern of co-dependence between journalists and sources. In particular, reporters are beneficiaries of information subsidy that enables them to file stories and meet deadlines. Sources are able, in turn, to insinuate their imprimatur on the news with the cover of a reporter's by-line. Moreover, as a question of habit and habitus, reporters are inclined to seek the conventional and legitimate authorities of the State and private sector to nourish their news narratives with information subsidy. Where the State is concerned, examination of news management tactics and strategies from, *e.g.*, the Reagan and Clinton teams, illuminate the relentless and often sophisticated attempts to seduce journalists. Finally, the Bush-era invasion of Iraq illustrates management of news media, particularly as it crystallizes in the practice of embedding, with its resultant sourcing patterns and news narratives.

Due to intensifying news production, "churnalism" and "McJournalism" have also been entrenched through the speed-up in the news factory. New media even heightens these problems by expanding the news hole, as documented by Fenton (2010a) and her colleagues. As always, there are important exceptions to the rule and the milieu of news is not predictably clockwork in its behaviour. Nevertheless, news production is also not an anarchic process and its systemic characteristics tend to privilege the status quo and established authorities for the reasons discussed in this chapter.

To summarize further, the three chapters that constitute this section have unpacked the infrastructure that surrounds news production; namely, private ownership, commercialism, sourcing patterns embedded within professional routines. Like concentric circles, each of these filters moves closer toward detailed discussion of the content of particular news discourses as they are subject to and illustrative of the Propaganda Model. Having considered context furnished by news infrastructure, Part II of this volume pivots toward analyses of text in greater depth as concerns dichotomized narratives, flak, and the rise of new media.

Notes

[1] As Bennett (2001) observes, the de-skilling of journalism has a long history and begins with the nineteenth-century advent of objective reporting (i.e., the rote emphasis on "who, what, when, where"). De-skilling also has numerous unexpected spinoffs. For instance, when news programs seek to raise their ratings, they do not double down on journalism and dig deeper into community affairs. Instead, "news doctors" make changes in theme music, studio sets, or in favor of soft or sensational news. One example: NBC made a much ballyhooed (and quickly abandoned) effort to raise ratings by introducing Connie Chung as co-anchor on its nightly news program in the 1990s. At the same time, the network shed reporters and slashed expenditure (Bennett 2001: 171).

[2] The language of autonomy that Phillips employs is borrowed from Bourdieu. Although he wrote little on journalism, Bourdieu (2005) proposes an autonomy/heteronomy binarism within its field: "the journalistic field is characterized, in comparison with the field of sociology (...) by a high degree of heteronomy. It is a very weakly autonomous field", which is to say it is strongly subject to influences from beyond it (2005: 33). Within this general situation of heteronomy, Bourdieu claims that journalists become differentiated "between those who are 'purest,' most independent of state power, political power, and economic power, and those who are most dependent on these powers (...)" (2005: 41). It is this contrast—in which some journalists are able to be more autonomous and to differentiate themselves —on which Phillips plays.

[3] Notice that Hayes & Guardino (2010) argue that the degree of media complicity with the Bush administration in the extended prelude to the invasion of Iraq was not as strong as has been assumed. Their evidence, however, pushes against even their carefully qualified claim.

[4] Along with absorbing Marine Corps manuals such as *Military Operations in Urbanized Terrain*, *The New York Times*' Andrew Jacobs comments as follows on his immersion in "one of the Pentagon's media boot camps" (Jacobs 2003: 1):

> All that marching, commiserating and drinking with the Marines makes for warm and fuzzy feelings on both sides...One memorable bonding episode was a snowball fight outside the barracks on our final day. (2003: 3)

Jacobs' report pirouettes between the "warm and fuzzy" posture toward the Marines and moments of self-reflexiveness about his participation in the corps' media outreach to enhance its profile. He finally posits, without any evident irony, that "American press coverage will serve as a foil to Iraqi propaganda or distorted reportage from less objective news outlets in the Middle East" (2003: 3).

PART II
Texts

CHAPTER 4
Weed Whackers and the Phantom Menace

The Bush administration's case against Iraq can be summed up in one sentence: Iraq has not led United Nation's inspectors to the weapons that Washington insists Baghdad is hiding.

—Michael Gordon, in *The New York Times* (2003b: 1)

There is only one truth, and therefore I shall tell you as I have on many occasions before, that Iraq has no weapons of mass destruction.

—Saddam Hussein "in an interview in Baghdad with a British Socialist politician, Tony Benn" (quoted in Preston & Weisman 2003a: 1)

While a small number of old, abandoned chemical munitions have been discovered, ISG (Iraq Survey Group) judges that Iraq unilaterally destroyed its undeclared chemical weapons stockpile in 1991. *There are no credible indications that Baghdad resumed production of chemical munitions thereafter, a policy ISG attributes to Baghdad's desire to see sanctions lifted or rendered ineffectual, or its fear of force should WMD be discovered.*

—Iraq Survey Group (original emphasis, n.d.: 1)

The New York Times reports that, in March 2012, in the San Diego suburb of El Cajon, Shaima Alawadi was murdered in her home (Lovett & Carless 2012). Death was inflicted by blows to the head from a tire iron (Rushe 2012). Thirty two years old at the time of the murder, Shaima was an Iraqi immigrant and mother of five. About a month earlier, a note had been taped to the family home's door that stated, "'This is my country. Go back to yours, terrorist'" (Lovett & Carless 2012). This (literally terrorizing) message was repeated in a similar note left beside Shaima as she was dying. Ironically, as the UK *Guardian* observed, Shaima's husband was a "defence contractor in San Diego, serving as a cultural advisor to train soldiers preparing for duty in the Middle East" (Rushe 2012: 3).

News accounts of Shaima's murder are, predictably, sombre and imply sympathy for the family since no newspaper would endorse threats and murder against an obviously innocent civilian; nor would any news organization take seriously the murderer's claims of "defence" against terrorism. Nevertheless, such a pattern was evident, indeed on a far wider scale, in the prelude to the invasion of Iraq in 2003. Repeatedly accused of being belligerent and terroristic, the nation of Iraq was invaded and subjected to external force in an action that was garnished with broad journalistic approval in the US. A raft of studies suggests that US news media recirculated government narratives that posited Iraq as a menace that necessitated armed confrontation (Boyd-Barrett 2004; Massing 2004; Moeller 2004; Scatamberlo-D'Annibale 2005). Hayes & Guardino (2010)'s extensive investigation emphasizes US journalists' search in the international community for oppositional voices; and it too furnishes considerable empirical support to the thesis of US news media capitulation to the invasion through extensive and affirmative citation of Bush administration officials.

In examining journalistic behaviour in the run-up to the 2003 invasion, this chapter trains the spotlight on *The New York Times*. The *Times* was selected as the case-study due to its lofty reputation—and for its *bête noir* status for the US' political right that mythologizes it as the "liberal media" nerve center. Notwithstanding expressions of violence-inflected malice from the political right toward the *Times* (discussed by Media Matters for America 2004), the newspaper's reputation for quality is such that it suffers no lack of modesty. The newspaper self-regards as "the world's greatest news-gathering organization (...) a colossus astride the globe"; "'To work at our newspaper is to experience awe,'" boasts one editor (Diamond 1993: 4). Curran posits the *Times* as best exemplifying "why the American model of responsible media capitalism has admirers around the world" (2011: 14). As the *Times* often showcases the pinnacle of US news tendencies, its discourses furnish a stringent test in evaluating the nation's news performance on an immensely important topic.

In crafting this analysis, articles from the *Times* were identified from a Lexis Nexis search using the terms "Iraq" and "United Nations" in the headline or leads from 1 January to 20 March 2003. One hundred and three articles, totalling 112,000 words, were thusly located. Beyond the *Times*' lofty reputation, it is also evident that the newspaper devoted considerable attention to the Iraq-UN news storyline. To wit, 37 of the articles identified in the Lexis Nexis search (36 percent) appeared on the front page of the nation's flagship newspaper.

In critically sifting through the *Times*' Iraq-UN discourse, one might claim that the newspaper should be given some slack. The view from hindsight is far less obstructed and "everyone" assumed in 2002-03 that Iraq was armed and aggressive. However, "everyone" did not assume such. In a journal article in 2002, I quote extensively from a former UN Special Commission (UNSCOM) inspector who had intimate knowledge of Iraq's capabilities from his extended engagement during the 1990s. In a 1999 interview, Scott Ritter stated that "from a qualitative standpoint, Iraq has been disarmed" (Goss 2002: 92). Ritter adds emphatically,

> When you ask the question, "Does Iraq possess militarily viable biological or chemical weapons," the answer is *no*! It is a resounding *NO*. Can Iraq produce today chemical weapons on a meaningful scale? No! Can Iraq produce biological weapons on a meaningful scale? No! Ballistic missiles? No! It is "no" across the board. (original emphasis, quoted in Goss 2002: 92)

Within the US military and intelligence communities, the assumption that Iraq had been effectively disarmed was also widespread

(Ricks 2007: 12-111). Professional mores regrettably inhibited members of this community from speaking out on the issue before the 2003 invasion.

Moreover, Iraq's actual state of immiseration under the regime of economic sanctions imposed after 1990 had been documented in grim detail by reputable sources by 2003. These sources include UN agencies (United Nations 1999) and their former employees (Ditmars 2002), academic investigators (Gordon 2005, originally published in 2002), investigative journalists (Pilger 2000), and the offices of US congresspeople (Sims, Zylman, Hickey, & LeClair 2000). Nevertheless, the extent to which Iraq had been ravaged by sanctions was not widely known given journalistic indifference and adherence to Clinton-era storylines on Iraq (Goss 2002). In contrast with US government and news media discourse, the reality of Iraq in 2003 was the country's devastated condition, *not* what dangers it allegedly posed to the global community.

Bush administration allegations about Iraq were repeated *ad nauseam* and then frequently re-circulated by news media that included the *Times*. The reasons that the news discourse played out as it did undoubtably include journalistic routines and sourcing patterns. Another factor in the tendentious and false accusations against Iraq was surely the habit of Us and Them dichotomization as a structuring logic that conditions news narratives. This is particularly true of foreign affairs in the news discourse of the US, a nation not only ensconced in its own peculiar ideological echo chamber, but situated between two oceans that buffer it from most of the world.

This chapter elaborates a theory of Us and Them in terms that enable application to news discourse. Thereafter, it furnishes the back story on the confrontation between the US and Iraq. Finally, and in detail, the chapter unpacks the filtering of reality through the *Times*' Us versus Them dichotomization. The take-away claim is that, in being strongly beholden to the Bush administration's Us/Them narrative structure, the *Times*' reportage pivoted decisively away from the facts on the ground.

Tools with Which to Excavate

Whereas Herman & Chomsky propose an anti-communism filter, this investigation employs a more abstracted Us/Them dichotomy in its stead. I posit Us/Them as a ready-made template for structuring the reportorial narrative. Retiring the anti-communism filter in favor of Us/Them is not due only to the sharp decline of self-proclaimed communist regimes after the 1980s. The Us/Them binary may also be exported across time and space as an instrument with which to analyze the discursive traces of fundamental conflicts within a society (and, as applicable, in a society's projections beyond its own borders). In this view, anti-communism is simply one exemplar of an Us/Them dichotomy that happened to drive a significant share of US news discourse from 1945 to 1990.

Why the emphasis on Us and Them? The binary could be called a primordial element within human societies. People are hardwired to be at once gregarious, in need of groups in which they may have the chance to flourish. At the same time, our species is often prickly in acting out a deeply grounded social reflex toward distinguishing between what is understood to be one's own group and what is *not* one's group. While it is glorious that we are different from each other, the enlightened objective of constructing Us is to achieve pride in the group, but not prejudice against other groups. In practice and under the sway of affirming Our sense of Us, They may be reified into being a coherent group even where none actually exists. For the purpose of solidifying the sense of Us, exaggerated and pejorative contrasts with Them may also be put in motion. Human discourse—in the vernacular of the street, in politics, in journalism—often does not enact the delicate calibration of pro-social pride that stops short of shading it into chauvinism.

In an extensive series of studies that analyze prejudicial discourse, Teun van Dijk characterizes the discursive moves of "polarization" or "Us/Them" as basic to the crafting of meaning in situations where conflict arises over group objectives and command over resources; which is to say, it arises in a wide swath of situations. Van Dijk writes, "Few semantic strategies in debates about Others are as prevalent as the expression of polarized cognitions, and the categorical division of people in ingroup (*Us*) and outgroup (*Them*)" (original emphasis; n.d.: pp.57-8). The differentiating move is partly realized "by attributing properties of *Us* and *Them* that are semantically each other's opposites" (original emphasis; n.d.: p.58).

Under the theoretically rich rubric of Orientalism, Edward Said's intellectual project more specifically addresses Us/Them binarization as it has played out between Occident and Orient. For Said, the West has to a large extent contrived a coherent identity by setting itself up in contradistinction to the East as its other, over which it had attained colonial mastery. While achievement of this identity corresponds with what Anderson would later call an "imagined community" (1983), the Orientalist project was not at all simple with concomitants palpably real and not imaginary. Said argues that Orientalism has implicated the "scientist, the scholar, the missionary, the trader" since the project was to "to manage—and even produce—the Orient politically, sociologically, militarily, ideologically, scientifically and imaginatively" (1979: 14, 3). Particular tropes that were mobilized by Orientalism's version of the Us/Them binary included the East as exotic, traditional, sensuous, mysterious—and brutal, untouched by the rigors of reason, subject to the primitive social organization of the leader and the led. These tropes continue to present an elaborate latticework of differences around a master Us/Them binary.

Importing the Us/Them binary into journalism studies, Prasun Sonwalkar (2005) argues that it is expressed mainly through sheer banality. In this view, the dichotomy recedes into familiar mood music and non-descript ideological wall paper: "'we-ness' is largely unstated and unarticulated," hovering "in the background as journalists go about performing their professional activities" (2005: 267). Channelling shades of Gans' classic analysis (2004), Sonwalkar posits that comforting we-ness is nourished by reporters' benign sense of their professional routines and decision making calculus. The world *just is* like that for the reporter, hence no further attention need be devoted to what is assumed about what is assumed. In these conditions, Us/Them binarizations are effortlessly produced within journalistic discourse as concerns, for example, socio-economic class, active dissent, or a nation's relations to the rest of the world.

Iraq Back Story

Before considering the *Times*' discourse, an overview of the transcript of the salient recent history is also in order. Four months before the invasion of Iraq, in November 2002, the United Nations Security Council passed Resolution 1441. The resolution placed the United Nations Monitoring, Verification and Inspection Commission (UNMOVIC) on the ground to assess Iraqi implementation of disarmament obligations and warned Iraq in the event of not adhering to the inspections regime. A "Joint Statement from the Popular Republic of China, the Federation of Russia and France" on 8 November 2002 underscored that Resolution 1441 "excludes any automaticity in the use of force" and notes statements from the US and UK delegations that suggested interpretive concurrence (Popular Republic of China, the Federation of Russia, and France 2002: 1). The three permanent members of the UN Security Council also state that, following reception of UNMOVIC reports, any action as concerns Iraq would be "for the Council to take a position." The interpretation in the Joint Statement was confirmed by a letter that US Secretary of State Colin Powell prepared for Syria's government that "'stressed that there is nothing in the resolution to allow it to be used as a pretext to launch a war on Iraq'" (quoted in O'Connell 2002: 2).

Shortly after the resolution passed, international law specialist Mary Ellen O'Connell concluded that "Resolution 1441 provides no new authorization for using force" against Iraq (2002: 1). After the fact, in a submission to the UK's "Iraq Inquiry," noted international legal scholar Phillipe Sands (2010) strongly endorses the widely understood parsing of Resolution 1441 that finds no trigger language in it and posits deployment of force would require a separate resolution. Despite notable consensus on the design and meaning of Resolution 1441, the *Times* largely accepted the Bush administration's contrarian, glib assertions that the resolution furnished all necessary legal cover for invading Iraq.

One-Two Punch

Disarmament is, of course, a noble goal and it should be demanded of other nations along with Iraq. However, in the often hysterical attention to Iraq's alleged weapons, of which there were none to speak of, the Bush administration effaced Iraq's actual condition by 2003. News media largely followed suit although the reality that surrounded Iraq was distinct from a nascent belligerent power. Comprehensive sanctions were imposed on Iraq in 1990, following Baghdad's invasion of Kuwait, and had not been lifted by 2003. The massive destruction of civilian infrastructure during the 1991 Gulf War (discussed in detail by former US Attorney General Ramsey Clarke [1992]) were not repaired under the terms of sanctions that endured until 2003 when new woes were visited upon Iraq.

Surveying the results of the Gulf War and sanctions as a walloping "one-two punch" makes for bracing reading. Mueller & Mueller, with appointments at University of Rochester and the School of Advanced Air Power Studies at Maxwell Air Force Base, claim that "If the U.N. estimates of human damage to Iraq are even roughly correct," sanctions caused "the deaths of more people in Iraq than have been slain by all so-called weapons of mass destruction throughout history" (1999: 51). Denis J. Halliday administered Iraq's shattered economy under the sanctions regime. Upon resigning his post as "Humanitarian Programme" coordinator, Halliday commented, "'I refused to continue to take Security Council orders, the same Security Council that had imposed and sustained genocidal sanctions on the innocent people of Iraq. I did not want to be complicit'" (Halliday 2003: 2). Halliday's successor, Hans von Sponeck, followed suit and resigned less than two years later for the same stated reasons (Everest 2001).

In more quantitative terms, the UN's own investigations into Security Council-imposed sanctions testify to the scale of social and economic destruction. Iraq had been "generally above the regional and developing country averages" on economic indices by the end of the 1980s. Health care reached 97 percent of urban residents, a "well developed water and sanitation system" was in place and education was characterized by "sizable investments" (United Nations 1999: 3, 4). In the immediate aftermath of the Gulf War in 1991, Iraqi GDP is estimated to have plummeted by two-thirds with even steeper drops in per capita income in the years that followed (United Nations 1999: 3). By the end of the 1990s, mortality for under-five-year-olds had more than tripled over ten years earlier. Moreover, "dietary energy supply had fallen from 3,120 to 1,093 kilocalories per capita/per day by 1994-95" and almost all children were impacted by malnutrition in some degree. Previously effective hospitals had not been repaired or maintained after the Gulf War, and potable water was far less available. As a result, lethal water-borne diseases "have now become part of the endemic pattern of the precarious health situation" (United Nations 1999: 5). Dramatically stepped-up social problems—delinquency, non-attendance at school, social and economic impoverishment, begging, prostitution, isolation and alienation from the outside world—characterized the siege conditions of the sanctions regime.

In sum, Iraq had been blown back from relative prosperity to being one of the world's most deprived nations in the decade that preceded the US' 2003 invasion. While Iraq's severe and widespread deprivation by 2003 are not entirely exclusive from it presenting a threat to other nations, this possibility had been characterized as nil by figures with deep knowledge of the country before the invasion (Arons 1999).

An Overview of the *Times* Discourse: Any "Good News"?

Does the *Times* ever break out of allegiance to US officialdom and the reliance upon the Us/Them binary? In its extended discourse, the *Times* regularly cited criticisms of the neoconservative project via appraisals that often emerged from abroad. On some occasions, cogent Iraqi criticisms are also reported in the *Times* without appearances of winking irony. For example, one article concludes with an Iraqi general claiming it would be easier, cheaper and surely more effective for the US to tell UNMOVIC where the supposed WMD caches were located to set up inspections, rather than lumbering into Iraq with an invading force of over 200,000 soldiers to try to seize them (MacFarquhar 2003b). The logic is irrefutable, if one really believes the weapons existed as asserted. As an occasional garnishing around the tendency to report in line with administration narratives, the *Times* also refers (briefly) to the devastating impact of the sanctions imposed on Iraq. The issue is covered in two paragraphs in one article (Preston 2003f), one paragraph in another (Santora 2003), and in one sentence in a third (Preston 2003a). The observations therein about Iraq's actual state in 2003—'"a very damaged place that needs enormous assistance"' (Preston 2003f: 10)—express the truth of the matter. However, in the midst of daily hystericism about Iraq's allegedly dangerous power projection, accounts of its material deprivation gained no traction as a sustained news narrative. Thus, the *Times* transmits blink-and-you-miss-it coverage of Iraq's actual state of pauperization under sanctions that contrasts with the constant and specious accusations on WMDs.

In the *Times*' extended discourse on the UN and Iraq in the run-up to the invasion, only one article is dedicated to digging a specific Bushian claim up by the roots. John F. Burns (2003) examines an Iraqi drone that, the reporter observes, had been touted as a fearsome weapon in a State Department fact sheet. He finds that the drone featured an engine "about powerful enough to drive a Weed Whacker," wooden propellers, and masking tape to connect its joints. The device had no payload, in contrast with State Department claims that it was weaponized, thus giving credence to Iraqi claims that the drone was for reconnaissance. Although any number of the contentious claims floated by Bushite neoconservatives could have been subjected to scrutiny—from regularly repeated tales of WMDs to the interpretation of Resolution 1441—Burns' report is the *Times*' clearest cross-examination of the charges that the administration trafficked.

"The Kindest, Bravest, Warmest, Most Wonderful Human Being I've Ever Known in My Life"

To lay out one further tool for analysis of the *Times*' news discourse, consider what Bennett positions as "the single most important flaw in the American news style" (2001: 35); namely, its orientation toward personalization. Bennett characterizes personalization as "the overwhelming tendency to downplay the big social, economic, or political picture in favour of the human trials, tragedies and triumphs that sit at the surface of events" (2001: 35). In Bennett's appraisal, personalization in the news is like a parody of Greek theatre as it channels melodrama that has been scrubbed of implications for the socio-political universe. At the same time, Bennett argues, it cultivates passive, fatalistic and depoliticized responses toward "larger moral and social issues" (2001: 35). Moreover, personalized news covers the sensational and the dramatic and not the quotidian with the monumental enfolded within it; the bus accident perpetrated by a bus driver asleep at the wheel, and not the ongoing state of the transit system on which the less affluent population depends.

While the *Times*' discourse on Iraq and the UN is not solely composed of personalization, it structures an appreciable share of the coverage alongside the Us/Them binary. This occurs mainly by construing the incipient invasion as showdown against Them, and specifically against the arch-villain Hussein. The regularity with which *Times*' reporters either use quotations or their own prose to describe Hussein in fiercely personalized terms recalls the fictional case of Raymond Shaw in *The Manchurian Candidate* (Dir: John Frankenheimer, 1962). Shaw's brainwashed comrades repeat the mantra, "Raymond Shaw is the kindest, bravest, warmest, most wonderful human being I've ever known in my life." The *Times* engages with similarly predictable phrases, albeit with a symmetrically unflattering valence. One guest editorial in the *Times* abundantly exemplifies the fixation with the Iraqi dictator as the anti-Raymond Shaw: Saddam Hussein "believes that by feigning cooperation (...)," "Saddam Hussein has no intention (...)," Saddam Hussein "will give us cooperation without doing anything to comply," "Saddam Hussein's rubber stamp legislature (...)," "Saddam Hussein is a liar (...)," "Saddam Hussein undoubtedly has more mock cooperation tricks up his sleeve," and so on, in the course of 1,315 Saddam Hussein fixated words (Indyk & Pollack 2003: 25). In this and a raft of other articles, via personalization, Hussein effectively *is* Iraq.

Aggrandizement of Bush, Our leader, also channels personalization. From the front page, Sanger & Shanker enthuse that, "President Bush is as determined as ever to move forward quickly and is not likely to be distracted by either logistical or diplomatic obstacles" (2003: 1). On 17 March, Sanger intensifies the visions of Our mighty leader at the helm: "President Bush and his supporters on the United Nations Security Council presented a stark choice to the deeply divided world body: Join a preventive war, or stand aside" (2003b: 12). Sanger doubles down in constructing Our leader as apotheosizing world-historic gravity with Churchillian accents:

"It is easily the most momentous decision of Mr. Bush's 26 months in office," "his fist punching the air for emphasis," "this is a new kind of war," "this is a war of 'liberation,' liberation of the Iraqi people," "the United States will not be threatened by weapons of mass destruction" asserted to exist because the neo-conservatives demand that it be so (2003b: 12). The personalization of Bush as a battle-hardened and wizened Raymond Shaw figure—embarrassing on their own terms when compared with the transcript of his life—are also symmetrical with the intensity of the scorn directed at Their tyrannical dictator. The paired exaltation and denigration of Our and Their leaders perhaps mutually summon each other into being. In this view, since Hussein is loathsome—a truism—necessarily Our figurehead must be a demigod. In any event, personalization underwrites the discourse and obscures the stakes.

Editorially Speaking

Analysis of the *Times*' editorial commentaries draws on both guest columns as well as unsigned editorials. To begin, it is important to notice that in its unsigned editorial of 9 March, the *Times* comes out against the incipient invasion. Nonetheless, the *Times*' formal opposition did little to halt editorializing and reporting that regularly transmitted the administration's premises and storylines that endowed the invasion with the "necessity" of confronting a recalcitrant Them.

In tracing the arc of the unsigned editorials, on 2 February, the *Times* exhibits what Neiger, Zanberg & Meyers (2010) have called "reaffirming criticism" that ratifies fundamental premises but ventures questions about tactical implementation. The editorial notes with exasperation that the Bush administration's rationales for invasion were changing on a regular basis and spokespeople were making claims (notably about nuclear arms) that went beyond the evidence. Nevertheless, the editorial beseeches the administration to make its public case more skilfully, to recruit Us into confronting Them. The *Times*' pleads that, "the United States needs to win the widest possible Security Council backing. That will require a more consistent—and convincing—articulation of the American case" (2003b: 14).

By 13 February, Bush's team had not fashioned what the *Times* desires: "a compelling case that he"—Hussein, of course—"poses an immediate danger to the vital interests of the United States" (*The New York Times*, 2003c: 40). The editorial also castigates France as it "must cease acting as if the real problem were to contain the United States." In fact, the US' unswerving drive toward aggression *was* the real problem at hand. Even where the *Times* wonders aloud about tertiary neoconservative claims, its discourse stays within the broad outlines that posit the US' incipient attack as a problem that necessarily originates with Their unacceptable behaviour.

By 9 March, and under the headline "No to War," the previously circumspect *Times* editorial line is more assertive in its criticisms and claims that UN inspections "could keep a permanent lid on Iraq's weapons programs" (2003d: 12). The editorial rejects the Bush administration's much trumpeted intelligence as "hunches." More broadly, the editorial laments the administration's strategic narrowness in not employing US influence to broker solutions to deep-seated problems in the Middle East (Israel-Palestine) while undermining UN authority for what the editorial characterizes as short-sighted objectives. Although the *Times* clearly departs from the administration in this editorial, the newspaper's discourses across its many sections regularly amplified neoconservative premises as will be further discussed below.

Guest Editorials

With respect to guest editorials, Adlai E. Stevenson III (2003) presents the most thoughtful peace advocacy. Stevenson argues that the Bush administration was employing double-standards about arms *vis-a-vis* Iraq and North Korea, needlessly inflaming the Muslim belt with prospective aggression, and ignoring the Israeli-Palestinian conflict; a triple play, Stevenson notes, hurtling toward a strategic dead end in a world with pressing problems to address.

However, in contrast with the urbane and restrained Stevenson, guest editorials in the *Times* generally pitched for the invasion with at times searing rhetoric. Kenneth M. Pollack was granted 1,315 words of editorial space on 27 January (with Martin Indyk) and another 1,700 words on 21 February for his solo editorializing.

Banging at neoconservative talking points, Indyk & Pollack warn repeatedly of what they characterize as a labyrinthine "inspections trap" set by the duplicitous Baghdad regime; the real trap in retrospect was the reality that Iraq, in fact, had nothing to hide. Pollack's histrionics reach greater pitch three weeks later. In the course of 1,700 words, he claims that Hussein's regime was "likely to acquire a nuclear weapon in the second half of this decade" (Pollack 2003: 27). In a personalized register, Pollack veers toward lengthy characterizations of Hussein's mental calculus that are crude even by the standards of the mind-reading genre. Pollack also steadfastly avoids factual engagement with the null findings of nuclear inspectors in the field in Iraq. In other words, warmed-over neoconservative memoranda with the attendant Hollywood "B"-movie versions of Them, find lodging on the *Times*' editorial page.

On 5 February, Barham A. Salih's guest editorial also advocates invasion albeit through appeals to Our humanitarianism. Described as "co-prime minister of the Kurdistan Regional Government in Iraq" that enjoyed *de facto* independence from Baghdad after 1991, Salih chimes at liberal bells in claiming that the invasion would be liberationist in outcome. He anticipates parallels with the endgame of World War II. Salih also makes pointed reference to Western protestors as misguided by what he implies to be simplistic analyses. The editorial unmistakably suggests that the marauding neoconservative invaders are better at being true liberals than the avowed (and hopelessly confused) liberal opposition.

An editorial by Ann Marie Slaughter breezily posits that, if after the fact, the invasion were to achieve its stated outcomes, there would be no outstanding legal or ethical issues. Identified as a dean at Princeton University, Slaughter advocates for invasion even if the UN process is blatantly subverted as a result. These glib arguments would have amounted to "might makes right makes the law, when We say it does" even if her assumptions had proven correct. As it stands, Slaughter's crystal ball of expectation ("irrefutable evidence" of WMDs, "the Iraqi people welcome their ["liberators"] coming") proved woefully wrong (2003: 33).

All Flavors of Reporting?

The *Times* presents the most sophisticated reporting available from US news media. The reputation is backed by the sheer volume of reportage on the Iraq/UN discourse. In one index of depth, an impressive 52 percent of the articles in the corpus were more than 1,000 words long and the overall average length was 1,087 words. The *Times*, on occasion, also publishes full documents that endow the reader with an unmediated gaze. For example, the Times published the full 5,000 word text of UNMOVIC chief Hans Blix's assessment of Iraq for the Security Council in late January (*The New York Times* 2003a). The *Times* also channelled many criticisms of the incipient invasion, mainly through frequently vigorous opposition at the UN. US officials were often depicted on the defensive in a world that was largely not buying its claims, with residues that regularly permeated the reporting.

Nevertheless, alongside concessions to reality, the *Time*'s discourse also copiously reproduced the premises of the Bush administration. In turn, dissenting commentary was positioned toward the edge of the discussion or as effectively contained in being quickly answered by administration themes and quotations marshalled by reporters. Furthermore, the "Threat and Responses" tag was applied to almost all of the newspaper's coverage in the corpus of articles. The phrase is an editorial intervention in itself, even—indeed, especially—when it appears alongside reporting. It signifies the *Time*'s summary judgement that the Iraqi Them was a real and not a contrived menace.

The *Times*' discourse is mainly straight—unbiased, objective—in what it reports. Michael R. Gordon's front page article of 18 January furnishes an example of reporting that weathervanes without apparent bias, spun around and back by competing claims. Gordon opens by noting that "the Bush administration has argued that the onus is on the Iraqis to (...) disarm" (Gordon 2003a: 1). These orthodox claims are quickly contested by UNMOVIC's Blix and a specialist at Washington's Brookings Institution. Gordon notes that weapons about which the Bush administration was making noise "have a range of about 12 miles", *i.e.*, they are limited use battlefield armaments and not existential threats to any nation. Although intricately balanced, Gordon's article begins and ends with administration claims that sandwich criticisms of those claims from less high-profile sources.

On many occasions, the news discourse's alignment with administration claims is subtle. For instance, reporters repeatedly cite Bush administration statements that there would be no "smoking gun" in weapons inspections of Iraq (*e.g.*, Landler & Cowell 2003; Weisman 2003a, 2003b). This administration meme, faithfully channelled by reporters, lowers expectations about substantial findings from inspections, even as administration neoconservatives were simultaneously polemicizing about Iraq's alleged arsenal. In other moments, alignment between *Times* reportage and the administration's claims is less subtle. Steven R. Weisman's front page report on Colin Powell's 5 February Security Council presentation reads administration assertions into the record under the rubric of objectively reporting what a newsworthy figure said: "Iraq's lethal weapons could be given at any time to terrorists who could use them against the United States or Europe," "'not an option, not in a post-September 11 world,'" "a second Security Council resolution authorizing force against Iraq would be desirable but not necessary," and so on (Weisman 2003d: 1). Besides transmitting Powell's claims with high fidelity, Weisman adds garnishing of approval around the "nearly encyclopaedic catalogue" of alleged evidence.

In another article, published on the front page, Felicity Barringer openly sides with administration claims. She quotes UN ambassador John Negroponte's assertion that the US "has ample legal authority to go to war under the previous resolutions" (Barringer 2003a: 1). Despite the speciousness of the claim, it is a proverbial "end of discussion" moment in Barringer's account. In the same article, Barringer writes that "Mr. Powell and Mr. Straw refused to be put on the defensive" by invasion opponents. British Foreign Secretary Jack Straw is described as having made a "frontal assault" on French foes, for which "he was greeted with applause" and later "described by one admiring Council diplomat" as displaying "'intricate wording'" (Barringer 2003a: 1). In such accounts, there is little doubt about who the righteous We are, as cued by the reporters' glowing descriptions.

While neoconservative-friendly stenography was a significant part of the *Times'* news mix, it was not the only tendency. James Risen (2003) both reports, and in some depth critiques, contentious claims made in Bush's 2003 State of the Union address. The *Times*' reporters can, moreover, be credited for having attended to bristling tension at the UN between the US and allies such as France and Germany. The reporting was, in these cases, driven by the fact of a genuine rupture among allies on which to report. By contrast, the expected domestic source of "legitimate controversies"—the opposition Democratic Party—failed miserably to confront the neoconservatives. In the handful of occasions when they are cited in the US-UN-Iraq discourse in the *Times*, Democratic politicians mint meagre reaffirming criticisms (*e.g.*, Preston 2003e), beyond the notable exception of Senator Robert Byrd (Stolberg 2003).

In part due to the general lack of an adversarial front from the opposition party, much of the reporting that filled the *Times*' pages was oriented toward contrived conflicts. For example, on 25 January, Julia Preston (2003d) follows her sources in reporting on a variety of clashes between UNMOVIC inspectors and Iraq over, for example, the disposition of "at least 11 documents" and helicopter paths through Iraqi airspace. Similarly, Ian Fisher frets about ground rules under which Iraqi scientists were to be interviewed by UNMOVIC (2003b). Given that the inspections were never completed due to the invasion, such details are at best "MacGuffins" that distract readers from the larger story arc of imminent US aggression.

Some reports hinted at the charade that was in motion. On page one, Preston writes: "The United States and Britain (...) continue preparations for a war in late February or March while demonstrating to sceptical allies, including France, that they are not rushing to judgement" (2003e: 1). As in this sideways admission, reporters generally played along with the US administration's ploys and did not call out the diplomatic pantomime even as they hinted at its essential phoniness. David E. Sanger & Thom Shanker blandly pass over the fact that the US and UK had already begun strafing Iraq with airborne attacks with civilian deaths by 3 March, more than two weeks before the ostensible onset of the invasion (Sanger & Shanker 2003). The bombing raid is reported in the twenty-fifth paragraph of Sanger & Shanker's 28-paragraph report that is otherwise concerned with the menace presented by (an already encircled and attacked) Iraq. This is all sensible when it is always already They who cause problems and who must be subject to Our chosen solutions.

Sourcing

As discussed in Chapter 3, sources have a privileged opportunity to insinuate their facts and interpretations into reportage. In considering sources, Michael Schudson insists that "after Vietnam," journalists' faith in the establishment centers of power "would never be the same, not even after September 11, 2001" (2011: 82). Schudson's assertion is indebted to fantastical yearnings about US news media being what it claims to be (i.e., unswervingly oppositional) and is innocent of any evidentiary backing. By contrast with Schudson's sentimentalized idealizations, in the corpus of articles, the *Times*' reporting focuses almost monolithically on sources from officialdom.[1] At the same time, only one article is devoted to *vox pop*. Not surprisingly, the *vox pop* quotations marshalled in Nagourney & Elder's (2003) report stay within the prevailing discourse on Iraq that had been rehearsed in mass media for years.

The more status that the source's position presents, the more he or she may be privileged within the discourse. Thom Shanker's report of 21 January 2003 presents vivid illustration of a mono-sourced article that largely repeats a series of claims from a single highly positioned official, Secretary of Defence Donald Rumsfeld. All of Rumsfeld's comments can be characterized as disastrously wrong in preface to the invasion: "'In the case of Iraq, we're nearing the end of the long road with every other option exhausted,'" Iraq "'is unique'" with regard to weapons programs, "'No other living dictators has shown the same deadly combination of capability and intent,'" and so on. As the reporter empowers Rumsfeld to drive the narrative in the direction of a risibly trite Hollywood script, the neoconservative line is reproduced with striking fidelity. In this case, deference is afforded to the authority of the speaker's position and not the merits of the claims.

Barringer and Gordon (2003) echo the administration's favored memes, albeit without mono-sourcing: "the lack of (Iraqi) action could be seen as lending credence to the American argument that Saddam Hussein has no intention of disarming," "Iraqi defiance would leave the pro-inspection arguments threadbare," "the [UN Security] Council was in danger of leaving itself irrelevant" (2003: 1). In phrases such as these, the reporters not only closely mimic the wording of the Bush administration, but they also accept the premise that *They* (Iraq) are the armed and dangerous party, no conditions attached.

Alan Cowell's 959 word report, datelined London on 17 February, gives the most extensive, if indirect, voice to the public in the discourse. Cowell writes that "at least 750,000" people had protested the incipient invasion in London over the previous weekend, while estimates for protest across Europe ranged from 3 to 6 million; "the continent has not seen protests on that scale in memory," he adds (2003: 10). Cowell also notes that the capitals of countries where right-leaning governments had lined up with Washington against their publics' wishes witnessed the largest protests (London, Madrid, Rome). While Cowell's reporting conveys the magnitude of public opposition, most of the *Times*' coverage of the issue is far more muted. For example, on 28 January, John Tagliabue writes with notable understatement that, "Public opinion across Europe is ranged against the war"—even before the heavily telegraphed "war" occurred (2003: 12).

Times writers repeatedly referred to the triad of US, UK and Spain as driving every gesture for the invasion of Iraq (*e.g.*, Sanger 2003b; Sanger & Hoge 2003). Nevertheless, in Spain where I reside, polls found a near monolithic 91 percent of respondents opposed to the invasion (Tremlett & Arie 2003). Numbers aside, spontaneous anti-invasion expressions during Spanish television programming and on banners hung from windows of apartments on seemingly every urban block ("*¡No a la Guerra!*") make for surreal comparison with the *Times*' version of Spain as an eager belligerent. In any event, the general effacement of popular dissent kept the lines of Us/Them conflict stark and largely unblurred.

Us and Them

Given the general characteristics of its discourse, it is not surprising that the *Times* also inserts Iraq into Orientalist tropes. In particular, they are mysterious, brutal, alien to truth and truth-telling, dangerous—even if they are *at the same time* childishly obvious and clownish. To be trapped within otherizing tropes is to be at once muscle-bound giants (menacing!) and simultaneously anaemic weaklings (bozos!). Moreover, in the Orientalist gaze, their social organization is simplistic and merely consists of the leader and the undifferentiated mass of subordinates who are led. These are all patently preposterous tropes when unpacked and laid out in this manner. Yet, they permeate almost all of the reporting on Iraq in the corpus of articles.

While the Us and Them framework may seem too simple for the *Times*, its news discourses on Iraq strongly exhibit such polarization. To take one example, Weisman's 23 January front page story is structured around Us and Them, augmented with accents of personalization. Weisman writes,

> Labelling him a "dangerous, dangerous man with dangerous, dangerous weapons," the president [Bush] said in Saint Louis that "if Saddam Hussein will not disarm, the United States of America and friends of freedom will disarm Saddam." (2003b: 1)

We *are* freedom. Ergo, Our friends are the friends of freedom—and there is correspondingly little doubt that Hussein, personalized index of Iraq, is Them.

To be Them is to be fake, and predictably so. The *Times* channels this otherizing logic when US officials "predict" with apparently unerring accuracy what Iraq will do next, even if (or because) those predictions may be more accurately construed as pre-emptive poisoning of the well of interpretation within news discourse. From the platform of a front page article, "Mr. Bush had predicted a token move by Mr. Hussein to alleviate pressure and divide the Security Council. He said, 'the Iraqi actions are propaganda, wrapped in a lie, inside a falsehood,'" lamina of subterfuge typical of Them (Barringer & Sanger 2003b: 1). National Security Advisor Condoleezza Rice is similarly elevated to the status of oracle in the *Times*: "'I can absolutely predict' (...) that Mr. Hussein will offer 'a little cooperation in hopes that he can release the pressure.' Then, she said, 'he goes back to cheating and retreating and deceiving again'" (Barringer & Sanger 2003a: 1). In a typical Us/Them moment, Stevenson & Sanger paraphrase Rumsfeld as he insists that, "The Iraqi government has proved very skilled in denial and deception tactics"—a cardinal feature of Them who do not reach Our rarefied levels of truth speaking (2003: 1).

However, given how sneaky Iraq is in Bush administration/*Times* discourse, the fact that there were "no smoking guns" of WMDs during the UNMOVIC inspections in 2002-03 *is itself the smoking gun.* When gazed upon through Orientalist spectacles, Their apparent innocence is only a confirmation of guilt and a penchant for disguise. Indictment is conviction, when We say it is. However, as noted in the epigrams at the start of the chapter, Hussein was telling the truth about telling the truth. Hence, as embarrassing as it is to notice, the Bush administration failed to reach the same standard of truth telling as a ragged dictator of a nation whose fortune's had sunk to calamity.

As a matter of definition, We are always already well meaning and correct. Therefore, "'The burden is upon Iraq'" to accede to whatever blustery demands or charges the US makes against it (Fisher 2003a: 1). At least in this particular case, Fisher's report allows a reasonable rejoinder from an Iraqi official about the presumption of guilt that surrounds the inspections (2003a). More typical is Landler and Cowell's dispatch on Iraq's "litany of failures and unanswered questions" as concerns its "cooperation with the inspectors" (2003: 1). Irreducibly and indivisibly, They are the problem in the first instance. Therefore, as demanded by the US administration, and as re-circulated in the *Times*, the task of satisfying Us is a Sisyphean one. They are simply not meant to succeed, as new demands replace already forgotten ones in an endless cascade that focuses relentlessly on Them, on Their misdeeds, on Their probity, on what They must (but by definition cannot) prove to Our satisfaction. It follows that, in the *Times*, Powell "insisted that the burden of proof was on Mr. Hussein, not the inspectors, to give an accounting of Baghdad's munitions," in what the *Times* celebrates as "muscular unapologetic" discourse from the Secretary of State (Landler & Cowell 2003: 1).

While it is usually not necessary to explicitly catalogue Our virtues when implicit comparisons via Their faults will do, Powell states that the immodest US' quest is to "'protect our people and to protect the world'" (Weisman 2003c: 1). The worst that comes of this quest is that some parties—for example, the UN Security Council (Zeller 2003)—may not immediately grasp the character of Our noble causes.

Other Thems

More ambiguous entities hover awkwardly around the frontier of the Us/Them binary and include the nation of France and UNMOVIC's lead inspector Blix. As concerns the former, Weisman (2003b) characterizes France as driven by naked economic self-interest in opposing the invasion of Iraq. France is also described as having mistreated ("'sand-bagged'") Powell. However, for the administration and the *Times*, the French are not so much menacing, like obdurate Iraq, as buffoonish. So it follows that, "an American official" muses with condescension that "'To the extent the French and others want to join us and catch up to us to keep the Security Council relevant—sure, why not?'" (Stevenson & Preston 2003: 1). Iraq is Them due to its depthless evil, while France is more of a wayward pet, on the penetratingly no-nonsense neoconservative view that the *Times* channels.

As for Blix, the *Times* largely constructs him and his UNMOVIC team's work according to its degree of alignment with the Bush administration's rhetorical needs of the moment. In January, Blix was inserted into the news narrative as a clear-visioned "tough guy" when his UNMOVIC team's reports were at their most sceptical (MacFarquhar 2003a; Preston 2003b). By March, however, the Swedish diplomat's claims that the work of inspecting Iraq could be completed within months presented an obstacle to the administration's designs on invasion. In light of the threat of Iraq being deemed a non-threat, coverage of Blix followed suit: "Much of the administration's frustration is directed not at Mr. Hussein, but at Mr. Blix," Weisman comments on 2 March (2003e: 13). Blix's stated desire for more time to complete the inspections, and likely declare Iraq in compliance, is reported to "have infuriated many in the Bush administration" (Weisman

2003e: 13). In these moments, the *Times* reportage on Blix' performance is congenial to the administration's objective of making an issue of the lead inspector's motives and judgement.

The *Times* also reports pressure on Blix to make judgements congruent with Our interests that was exerted by Powell (Cushman & Weisman 2003) and by Rice (Preston & Weisman 2003b). The *Times* closely reproduces the double standards embedded in an Us/Them binary in these moments. In contrast with the indignation that greeted Iraqi complaints about the conduct of inspections in January (MacFarquhar 2003a), the matter-of-fact manner in which the *Times* reports Bush administration public pressure campaigns suggests that its reporters find nothing curious in the US' exertions against the investigator's independence.

They Might Be (Existentially Threatening) Giants

While France was derided in the US' wider discourse as part of the "axis of weasel," the construction of recondite evil emanating from Iraq tended to conjure still more Thems. In particular, Hussein and Iraq's depravity was underscored through references to al Qaeda in a number of articles.

The neoconservative campaign to link Iraq and 11 September was emotively potent for having been launched only eighteen months after the massacres. It was also risible mass manipulation, particularly in the light of Clinton-Bush-era anti-terrorism coordinator Richard Clarke's book-length argument that Bush's team was complacent about al Qaeda in the months preceding 11 September (Clarke 2004). Amplifying the neo-conservative pre-invasion mood music, Chief of Staff Andrew Card intoned that "'the United States will use whatever means necessary to protect us and the world from a holocaust'" (Landler & Cowell 2003: 1). Days later, Bush postured gravely in a front page report: "'After September the eleventh, the doctrine of containment just doesn't hold any water'" (Stevenson & Preston 2003: 1). In turn, the *Times* reporters regularly incorporate administration cues to articulate Iraq to al Qaeda.

David E. Sanger's (2003a) front page article on 30 January presents one means by which the Iraq-al Qaeda articulation was contrived. Sanger claims that the US has ample evidence of Iraq-al

Qaeda ties, but must be cautious in presenting it as to not compromise sources and methods. In this double bind, Their evidence (Iraqi compliance with disarmament goals) is summarily dismissed as meaningless pantomime. At the same time, evidence too secret to be discussed (of Iraq-Al Qaeda collaboration) is also of unquestioned veracity since it comes from Us. Despite the hair-on-fire nature of the alleged Al Qaeda links to Iraq, little journalistic investigation of the claims appears in the *Times*. One exception is a largely inconclusive, *Roshomon*-like story on Abu Mussab al-Zarqawi's relation to Iraq (Van Natta & Johnston 2003).

In what Stevenson & Sanger call a "spirited news conference," Rumsfeld thunders, "'In the case of Iraq, the task is to connect the dots before there's a smoking gun. (...) If there's a smoking gun, and it involves weapons of mass destruction, it is a lot of people dead, not 3,000 but multiples of that'" (Stevenson & Sanger 2003: 1). Stevenson & Sanger's article finishes with this chilling observation, a literal last word. Rumsfeld invokes scenes so horrific that there may even be lingering reluctance to point out that there continues to be no evidence that Iraq had the will or means to carry out such attacks. Despite the many opportunities to present independence from the government in its coverage, the *Times* conveyed Rumsfeld's unseemly visions and reckless screeds intact, following the neoconservative plotline about the nature of Us and Them.

Stepping back to survey the larger picture, one may also see that Iraq's economy had not only been smashed by more than a decade of sanctions. By early 2003, Iraq was also surrounded by more than 200,000 US forces from the world's most extravagantly armed war machine. Moreover, as noted in the *Times*, bombing raids had already been conducted by early March. Yet, in administration accounts that the *Times* largely followed, *Iraq was menacing the US*. In this view, the circular logic of Us and Them demands that Their depravity has no recognizable limits. Indeed, no recognizable logic can even ground claims about it.

Along with conjuring apocalyptic scenes of the future, the recent past is also re-arranged as needed. The *Times* frequently quotes US officials, from Bush on down, as they assert that passage of UN Resolution 1441 endowed the US with legal authority from the Security Council to invade Iraq. As discussed earlier, the spirit and letter of the resolution were designed to set up a "two-step" process if there was to be armed intervention in Iraq; this would, in turn, demand evidence that Iraq posed actual danger. The two-step design was instrumental to the unanimous passage of the resolution by the Council.

While the *Times* does not examine the issue by, for example, consulting international law professors, the newspaper's discourse repeatedly quotes US officials asserting the legality of an invasion based on Resolution 1441. Bush is thusly quoted by Stevenson & Preston (2003) while badgering the Security Council about the resolution that "provided *him* with all the authorization *he* needed to take military action" (emphasis added; Stevenson 2003: 1). Similarly, Preston assures readers that, "Bush administration officials argue that the resolution gives the authority to attack Iraq if the Council declines to act" (2003c: 8)—a manifest absurdity even on its face in enabling simultaneous and mutually exclusive US and Security Council interpretations of the same resolution (see O'Connell 2002). Other reports in the *Times* claim, for example, that "'the legal authority [to invade] is clear without a vote'" on a second resolution, even as reporters' subsequently place some muddle around the issue (Sanger & Hoge 2003: 1). Despite the magnitude of events, the *Times* does not engage with a direct forensic examination of the administration's regularly minted assertions on Resolution 1441; what the neoconservative We says is enabled to define the contours of reality, however speciously and to whatever disastrous effects.

Conclusion

Despite the *Times*' status as the pinnacle of US journalism and its place in the popular imagination as a left-leaning news source, the newspaper's extensive reporting and editorials in early 2003 regularly dovetailed with the Bush administration's narratives and priorities. Even as Iraq was manifestly a country in shattered condition after more than a decade of sanctions, it was discoursed upon as a Them that posed an unavoidable problem for its imminent and hostile power projections. Moreover, Iraq was regularly personalized as a monolithic efflux of the malign character of Hussein.

The qualitative analysis presented in this chapter furnishes evidence that Us and Them dichotomization provided the structure that organized the news narratives and occluded salient features of reality. The next chapter will similarly unpack a news discourse as refracted through the Us/Them binary. In this case, however, the dichotomization filter will be applied to a domestic story from the United Kingdom in an effort to export the model outside the US news discourse to which Herman & Chomsky dedicated their focus. While there are hazards in exporting models of journalistic critique beyond national borders (Curran 2011: 7-46), one stake in this discussion is that the retooled Us/Them filter may be made-to-order for the methodological purposes of news analysis across different societies.

Notes

[1] Five articles were randomly selected from the corpus to examine the claim that the discourse stayed tightly within elite sourcing patterns (Barringer 2003b; Bumiller 2003; Preston 2003b; Preston & Weisman 2003a; and Wines 2003). Across almost 5,200 words, 35 different sources, and 75 quotations and paraphrases, all of the sources cited in the random sample (100 percent) can be straightforwardly characterized as from the rarified precincts of officialdom. Quoted and paraphrased sources included figures from the US government (*e.g.*, Bush, cited 10 times in Bumiller 2003), UN agencies (Blix, cited 6 times in Preston & Weisman 2003a), figures from governments outside the US (Germany's UN ambassador, cited twice in Barringer 2003), and think tank scholars (Alexander G. Rahr, cited 3 times in Wines 2003).

CHAPTER 5
Feral Peril: Broadsheets and the British Street

This chapter's investigation expands the Herman & Chomsky model in several notable ways. Specifically, their proposed "Anti-Communism" filter has, in a post-Soviet era, been reworked into a more flexible "Us and Them" filter that structures many news narratives. For being more abstracted into Us and Them, the refashioned filter exports across time and space to better fit the specifics of news performance of particular societies at given moments. On the assumption that all societies harbor their own versions of Us/Them dichotomies, the remit of the original Propaganda Model is made more supple. In this chapter, the Propaganda Model is also taken "off shore" in a further departure from Herman & Chomsky's concern with US news media. In particular, this chapter investigates the 2011 London riot coverage in *The Guardian* and *The Daily Telegraph*, British broadsheet newspapers of quality with respectively left- and right-leaning editorial lines.

The application of the Us and Them filter in this chapter presents a further shift with respect to the original Herman & Chomsky analysis. I will argue that *The Guardian* eschewed the temptations of the Us and Them binary as it reported and informed about the riots. By contrast, sustained instances of positively appraised journalistic performance are rare in Herman & Chomsky's discussion. Of further moment is that *The Guardian*'s indifference to Us/Them dichotomizing may be grounded in its partial buffering from commercial demands. The newspaper may thus serve as an illustration of what can happen with lighter commercial pressures exerted on news media. In this view, *The Guardian* presents a case study that is highly salient to the implications of the Propaganda Model—even as (or because) it does not behave in close accordance with the model for reasons that the proposed filtering mechanisms may at least partly explain.

To demonstrate the differences between *The Guardian* and the *Telegraph* in their riots discourse, I will survey some general tendencies in their reportage. Thereafter, I will telescope in on their contrasting editorial lines, coverage of police, Member of Parliament David Lammy, and the "Reading the Riots" project. Before the turn toward the two newspapers' content, I furnish back story on the riots and on the two contrasting newspapers' politics and conditions of production.

The Riots: Proximal Events

On the last day of 2011, the London-based *Guardian* presented a 4,100-word round-up of the year's news stories (Aitkenhead 2011). Although 2011 was an eventful year, it is not a surprise that the London riots were the lead item in Decca Aitkenhead's account. The toll of the mass disturbances was estimated at five dead, and 4,000 arrested out of an approximately 13-15,000 people who took part between 6 and 10 August 2011. Moreover, the cost of the resultant damage was estimated at £500 million (Davies 2011).

The explosive unrest was prefaced by the police's fatal shooting of 29-year-old Mark Duggan on 4 August 2011 by Scotland Yard's gun crime unit in the London neighborhood of Tottenham. In the immediate aftermath of Duggan's death, the police claimed that the suspect had shot first, a story that was later recanted. Adding insult to lethal injury, the police did not promptly communicate with Duggan's family. As a number of members of minority group members have died during brushes with police or in custody over the years in London, protests quickly followed. Protest mutated into rioting in London and other English cities that witnessed arson, looting, and crimes against persons. In one particularly heinous incident in Birmingham, a car lethally rammed three people who were guarding a car wash from looters. After flooding the streets with 16,000 police, the riots were quelled by 11 August. In the immediate aftermath, guidelines for sentencing were severe by design even for clearly peripheral participants in the events.

Introducing the Players

The 336 articles that were examined in this paired case study were gathered through the Lexis Nexis database. The parameters for the search of the two newspapers were "London" and "riots" both contained in an article's headline or lead for dates between 10 August and 31 December 2011. The resultant corpus consisted of 147 articles from the *Telegraph* and 189 from *The Guardian*. Hence, 29 percent more articles were identified in *The Guardian* while using the same search parameters. The difference in the two newspapers volume of coverage is, on closer examination, appreciably greater. The average riots-related story length in *The Guardian* was 762 words, whereas in the *Telegraph* the mean was 449 words. Moreover, the sum of words published in *The Guardian* (144,000) more than doubled that of the *Telegraph* (66,000). I will argue that *The Guardian* not only devoted more coverage to the riots, but was far more compelling in capturing the texture of truth than its right-leaning competitor.

In covering the riots, *The Guardian* hews to characteristically left-of-center concerns and interpretive strategies. These include emphasis on social context over monadic individualism. Although it exhibits pluralism, as when it sought out reactions to the riots from a panoply of different actors, *The Guardian* regularly cross-examines the neoliberal "free market"'s promotion to the status of steering mechanism for society by placing it in the back story of the riots in a number of moments. The *Telegraph*'s editorial line is, by contrast, right of center and supportive of the Conservative Party. The newspaper "'Has the reputation of being the newspaper of the establishment'" and "'even the paper of record over *The Times*'" (BBC News 2004). The *Telegraph* was named Britain's "Paper of the Year" in 2010 largely in recognition of its coverage of expense abuses by members of Parliament (*The Daily Telegraph* 2010). As one BBC correspondent comments, the *Telegraph* is "'a steady supporter of a very traditional way of life which chimes very much with the Tory party'" while "'its political editors have always been seen as having a hotline to senior ministers'" (BBC News 2004: 2). The *Telegraph* also imbibes the right's paradoxical ideological cocktail. To wit, the newspaper valorizes the nation and tradition, exalts neoliberal free markets (that regularly undermine national interest and tradition), and hails law-and-order while it expresses wariness of the State (the same State that maintains law-and-order and implements neoliberalism). There is further irony in the fact that the law-and-order *Telegraph*'s proprietor, The Lord Conrad Black of Crossharbour, was convicted in 2007 of defrauding the newspaper's parent company of $(US)6.1 million (*The New York Times* 2011).

Among quality "broadsheet" papers in Britain, *The Daily Telegraph* is the circulation leader with 894,000 copies sold per diem (Franklin 2008: 8). By contrast, *The Guardian*'s daily circulation of 371,000 exceeds only the *Independent* among the broadsheets. Although the *Telegraph* was the first British paper to move online, precociously in 1994, by 2006 *The Guardian* boasted "the largest dedicated internet staff amongst the national newspapers" with 60 journalists (Lewis, Williams, Franklin, Thomas & Mosdell 2006: 10). While both are among the most visited online news sites in the UK, *The Guardian* pulled ahead of the *Telegraph* in online readership by March 2009 (Kiss 2009).

Revisiting the Ownership Filter

The Guardian's relation to commercial media is of interest. On one hand, the newspaper is partly subsidized via the Scott Trust. On the other, along with being sold at the newsstand and via subscription, it is subject to demands to attract advertising. Indeed, in 2006, *The Guardian* was (literally) the biggest British national daily newspaper with a per diem average of 89.3 pages per print issue—and it averaged highest among its peers in ad content at 31.6 pages per day. By comparison, the *Telegraph*'s 2006 averages stood at 66.7 pages per day with 21.3 devoted to ads (Lewis, Williams, Franklin, Thomas & Mosdell 2006: 11).

Lewis, Williams, Franklin, Thomas & Mosdell (2006) explain *The Guardian*'s strategic foresightedness in terms that put the ownership model into the frame:

> The unique aims of the Scott Trust (...) mean that Guardian Newspapers Ltd are not subject to the same short-term economic pressures as compa-

> nies which have to report to shareholders, and they have consequently been able to invest in the long-term future of their internet news portal *Guardian Unlimited* while absorbing some fairly substantial losses. Indeed, it is worth noting that by far the most successful British online news providers—the BBC and the *Guardian*—have both had their operation subsidized. This may indicate the commercial limits of online news sources. (2006: 25)

In light of the subsidy and the vast ad accounts, *The Guardian* is a commercial and subsidized hybrid. It is significant, in light of earlier discussion in Chapters 1 and 2, that even partial removal from commercial pressures is associated with noticeable differences in the resultant news discourse. Lewis, *et al.* report that, while all British broadsheets regularly use pre-packaged materials from wire services and PR, *The Guardian* "is the most independent of the newspapers" (Lewis, Williams, Franklin, Thomas & Mosdell 2006: 25). While two-thirds of the *Times* and *Telegraph* reportage is partly or wholly grounded in pre-packaged materials, half of *The Guardian*'s articles were generated by enterprising staff. Moreover, within the half of *The Guardian*'s reporting that was derived from pre-packaged (informationally subsidized) materials, Lewis and colleagues conclude that it was more likely than competitors to make substantial transformations of those materials. While this does not demonstrate that *The Guardian* is cut from wholly different cloth, it suggests noticeable differences as concerns journalistic practice.

Overview of the Riots Discourse in the Two Newspapers

As the *Telegraph* and *The Guardian* are covering the same set of events, it is given that the two newspapers report some of the same riot-related material. For example, both cover some (but not much) international reaction (*Guardian*: Khaleedi 2001, *Telegraph*: Beckford, Hughes, Harding & Hough 2011). Differences in coverage are, nonetheless, profound. Distinctions between the two papers arise even in the contrasting results when their reporters took to the streets to report. In a 2,920-word account, *The Guardian*'s Paul Lewis (2011a) donned a "hoody" and aroused suspiciously inflected requests for cigarettes while positioned at the riots' frontlines. He was otherwise able to report as he witnessed shocking scenes of destruction. By contrast, the *Telegraph*'s Peter Hutchinson's bulletin from the street does not proceed further than having been "dragged to the ground, held in a headlock, punched in the jaw and mugged" (2011: 3). Discussion of gangs was also distinct in the two papers. While no one outside of a gang supports them, for obvious reasons, the *Telegraph* flogged at them as a convenient scapegoat for the riots. *The Guardian*, by contrast, reported the police estimate that 20-percent of rioters were gang-affiliated (Laville, Dodd & Carter 2011). While this figure is non-trivial, it also means that four in every five rioters were *not* in gangs and that facile finger pointing toward Them is not adequate to the observable facts. While the *Telegraph* loudly, but not quite accurately denounced the role of gangs, *The Guardian* also reported on community youth workers with dramatically reduced budgets in the Cameron austerity era who assay to keep young people from joining gangs (Williams 2011).

The Guardian's coverage is not simply more than that of the *Telegraph*; it is superior in informing its audience while generating multiperspectival news from its left-leaning anchorage. To take one example, Patrick Wintour and Paul Lewis conduct an interview sketch with Iain Duncan Smith, the Conservative Party's Social Justice Minister. In addressing the riots, Duncan Smith summons a brigade of clichés: "'so many families are broken,'" "'We seem to be a society that celebrates all the wrong people,'" "'a sense of structure and authority in kids' lives had collapsed,'" while Duncan Smith claims that threat of arrest is dismissed on the street as "'a joke'" (quoted in Wintour & Lewis 2011: 1). Although the dysfunctions on which Duncan Smith fixates are problems of moment, the minister arguably privileges individual pathology as the clue that solves all crimes. Moreover, *The Guardian*'s sketch of Duncan Smith presents a concern with reporting what the paper's political foes in the Conservative Party think in reasonably dispassionate terms. It is a display of reportorial curiosity that the *Telegraph* never exhibits toward any of the political actors whom it rigidly constructs to play the part of Them.

In stark contrast, the *Telegraph* invites the criticism that it posits knowledge as preceding knowing and that it is nakedly ideological in the first instance. The *Telegraph*'s Phillip Johnston expresses this view with gusto: "To find out what has gone wrong," as concerns the riots, "we do not need to delve too deeply into the specific causes of the appalling events of the past few days, or establish commissions and inquiries" (2011a: 19). In this view, arm chair assertion and lazily crafted Us/Them dichotomies will suffice.

Editorials: *The Daily Telegraph*

The right-leaning *Telegraph* invests heavily in Us/Them binaries to structure its news, with associated enthusiasm for the efficacy of repressive force. As the riots were reaching endgame on 10 August, Johnston discusses Them in terms that he posits as "on the lips of all decent people," who constitute Us (2011a: 19). Gazing through classist spectacles on "cities [that] were trashed by elements of their criminally inclined underclass," Johnston concludes that They demonstrate themselves to be "feckless, mindless and amoral thugs," disgruntled monads driven by "perceived grievance" (2011a: 19).

In an unsigned editorial, the newspaper attacks Home Secretary Theresa May from her right flank for having "ruled out the use of water cannon and the deployment of the military to help quell the disturbances" (*The Daily Telegraph* 2011a: 21). According to the *Telegraph* editorial, her "civics lesson has clearly been lost on the thousands of young thugs who have laid waste streets across the capital (...) with brazen impunity" (2011a: 21). In the *Telegraph*'s vision of civic schooling, "the thugs must be taught to respect the law the hard way" via tear gas and rubber bullets. Again exhibiting its faith in the repressive dimensions of State authority and a penchant for hard sticks over carrots, the *Telegraph* recommends "exemplary sentencing" for anyone convicted.

Beyond its call for the extraordinary measure of military intervention within Britain's borders, the *Telegraph* claims that police have "become so sensitized to the issue of race post-Macpherson that their response to violent criminality when perpetrated by predominantly black people can be fatally inhibited" (2011a: 21). Johnston similarly blames "four decades of politically correct policing" for the turmoil, thus extending the *Telegraph*'s penchant for glib caricature that invites the inference that police had been patrolling the streets with hand-wringing and fumbling apologies. While the *Telegraph* otherwise says little about race in its riots discourse—and even confronts chauvinistic *vox pop* commentary in its reporting (Gilligan 2011)—the riots lubricate occasional hoary asides about Them in racialized terms that include the coded phrase "political correctness."

Aside from personalizing the riots as mindless bad behavior, the *Telegraph*'s solutions are similarly personalized via faith in top-down authority exercised by an aristocratic, "born to rule" politician. In contrast with the prolix Home Secretary, the *Telegraph* approvingly editorialized that Prime Minister David Cameron "returned early from his holiday to take control of the crisis" and "set a new and aggressive tone" (2011a: 21). Johnston similarly enthuses over Cameron's instincts toward "robust police response and condign punishment" as the PM is characterized as comprehending that his "first duty is to keep order and protect property" in a distinctly crabbed concept of the State (2011a: 19). By contrast with Cameron's visions of the State as punisher, former London mayor Ken Livingstone claimed that massive budget cuts cued the mood music for the riots—and is editorially dismissed as "grotesque." The *Telegraph* thereby places "Left-wing politicians" among the deviant Them as enablers (Johnston 2011a: 19).

Aside from some mild efforts to understand the riots as morally confused (Crompton 2011), and nods toward the socio-economic climate from a less conventional Conservative London mayor soon to be seeking re-election (Boris Johnson 2011), the editorial response in the *Telegraph* plays heavily on Us/Them. Moreover, the *Telegraph* emphatically invigorates the neoliberal State that massively shifts its functions from the social democratic (the school, the library, the social worker) to the coercive (the police, the military, the prison).

Welcoming "Borat Values"

Despite fierce Us/Them dichotomizing, the *Telegraph* avoids otherizing in one particular direction. Britain's immigrant communities are singled out for praise in several opinion pieces (Archer 2011; Daniel Johnson 2011; Odone 2011). For example, Christine Odone is impressed by "Turbaned Sikh men [who] stood guard outside their temples last Tuesday night. Some held swords, others hockey sticks as they defied looters" (2011: 20). Although she later adds a wince-inducing reference to "Borat values," Odone sentimentalizes the immigrants' "tight knit enclaves" and what she takes as immigrant devotion to tradition: "divorce is almost nil, single motherhood ditto" (2011: 20). Daniel Johnson (2011) similarly writes with apparent feeling about three young men from an immigrant family who were killed while defending a car wash business from looters in Birmingham.

Alongside idealization of the immigrants, each of the pro-immigrant *Telegraph* editorialists pivots to otherize in another direction. Following his celebration of immigrants, Graeme Archer rants: "This is de-civilisation. This is what happens when middle class liberals suspend judgment, for fear of causing offense" (2011: 20). In this discourse, acceptance of immigrants is exchanged as an argumentative token in order to otherize "middle class liberals" who are nonsensically asserted to cause mass rioting through their alleged reticence. Does Archer *really* believe that the riots would have been prevented by a concerted campaign of insults? Similarly, in Johnson's editorial, a family's ineffable loss quickly becomes a set piece in which to stage otherizations: "If our welfare and education systems had not created an underclass of feral youths, Haroon [Jahan] and his friends would not have been the victims of mob violence." Johnson cites no evidence and furnishes no logic that could hold his assertions of blame in alignment in what collapses into a neoliberal canned speech. Nevertheless, he pirouettes effortlessly from the anti-Statism that indicts the educational and welfare systems of murder to steadfast Statism: "For the first time, Londoners are learning what it feels like to be properly policed," Johnson enthuses in finding the rainbow after the storm (2011: 19). While the discourse on immigrants is notably sentimentalized, even patronizing, it enables the *Telegraph*'s Us/Them terms to be mapped onto the left-leaning political opposition.

Editorials: *The Guardian*

The Guardian conjures a different world in its editorials from that of the *Telegraph*. In characterizing the difference, a *Guardian* editorial from 29 August is illustrative in its explicit rejection of Us/Them binarizing. The editorial begins by problematicizing George W. Bush's "with us or against us" mode of dichotomization. In the wake of the riots,

> (...) echoes of Dubya's menacing rhetorical contrast have been heard much more widely across the policy front. And it is not only individuals but whole families and even entire communities, which the right reckons have now been revealed as so indelibly wicked that the country can simply forget about the obligations it previously owed them. (*The Guardian* 2011d: 26)

As concerns Conservative Party plans to evict families of rioters from publically subsidized council housing, the *Guardian* editorial objects on the practical grounds that people thusly expelled will be in dire straits and stripped of further stakes in society. However, "an even stronger objection is the failure of this underclass-baiting to respect the separateness of persons." Sanction of a whole family for one member's behavior squares with collective punishment that characterizes other forms of prejudicial abuse under the pretext that They are, after all, an undifferentiated Them.

The Guardian's editorial line was not only distinct from that of *Telegraph* as the riots played out, but manifested its difference from early moments in the discourse. An editorial on 10 August takes *The Guardian*'s hardest line on law and order, even as it develops a carefully measured understanding of events:

> (...) important caveats notwithstanding, there is only one right side to be on. The attacks, the destruction, the criminality and the reign of fear must be stopped. The rule of law in the cities of Britain must not only be defended against delinquent destruction. It must also be enforced. There can be arguments about wider issues later. Today, in the moment of threat, the necessary position is to stand behind the police. (*The Guardian* 2011b: 28)

In other words, in lieu of baseline security, there is no community. The same editorial also flies its left-of-center colors in commending the government for resisting calls for steps toward militarization. Later, *The Guardian*'s editorial pages examine the "caveats" and "arguments" to which the 10 August editorial alludes. In the immediate aftermath of the horrifying events, on 13 August, *The Guardian* posits that "even criminality has to be understood, so that it can be prevented if possible" (*The Guardian* 2011c: 40).

On 11 August, *Guardian* columnist Seamus Milne (2011) makes further editorial moves to confront otherization of the sort that animated the *Telegraph.* Milne contrasts reactions from Cameron and former London Mayor Ken Livingstone. Whereas Cameron emphasizes Their "'criminality, pure and simple,'" Livingstone made a link between the riots and government spending cuts. Among some pundits, Livingstone's comments were received as if "he'd torched a building himself" (Milne 2011: 31). In this view, exemplified by *Telegraph* editorializing, "there was nothing to explain" in the rioting "and the only response should be plastic bullets, water cannon, and troops on the streets" (2011: 31). In the immediate aftermath of wrenching destruction, Milne posits the ostentatiously "common sense" answer of right-wing press and politicians as also presenting the "nonsensical position." He observes that riots had occurred at a moment when youth service budgets were being slashed—and *not* at some other moment. Moreover, the editorialist writes that the police practice of "stop and search" is 26 times more likely to be practiced on blacks who, in turn, manifested high participation in the riots. Finally, Milne confronts one of the right's indignant talking points by stating that concern for victimized small business owners is "fatuous" without a corresponding concern with preventive measures against future riots via examination of causes. From a less specifically British perspective in *The Guardian*, noted critic of neoliberalism Naomi Klein (2011) pulls further on some of these threads and construes the riots as enraged acting out against constricting opportunities for class promotion.

Academic researchers Jacopo Ponticelli & Hans-Joachim Voth (2011)'s editorial page intervention puts further rigor behind these arguments. Ponticelli & Voth's study of almost 40 nations across approximately 80 years found a strong statistical link between budget cuts and rioting. Moreover, the effect of the first variable is strongly amplified on the second. To wit, a budget cut of three percent intensifies into *100 percent* more indices of public unrest. While the researchers' language is technocratic, the message is a riposte at the right wing narrative of individual criminality bracketed off from its surrounding political economy.[1]

Police on Their Backs: *The Daily Telegraph*

Discourse on the police is a topic that strongly differentiates the two newspapers. Whereas *The Guardian* acknowledges the need for police to keep order, they also hold their conduct to the standard that the gravity of their work demands. Police are not so much an "Us" or a "Them" for *The Guardian* but an important actor and one that the paper regularly consults in sourcing its news reports. By contrast, *The Daily Telegraph*'s posture toward the police assumes that they must be valorized as one of "Us" and that their deployment of force necessarily generates desirable outcomes (and, conversely, withholding force unleashes or intensifies chaos).

The *Telegraph*'s support for police is robust as evidenced by one article that effectively recruits for the Metropolitan Police Service (Hoare 2011). In idealizing terms on the editorial page, the newspaper posits that police

> have been accused, among other things, of being too robust, of racism, of harassing ethnic minority youngsters and of swinging the lead. There are more of them than ever, yet fewer are seen on the streets. But that is because they spend so much time either filling in the forms that Parliament has foisted on them or carrying out duties that most people—and many police officers—do not think it is their job to perform. For decades, they have been encouraged by politicians, especially on the Left, to be another arm of the social services. (*The Daily Telegraph* 2011c)

None of the rapid-fire assertions are backed with evidence. Instead, the editorial summons Thems who are curiously powerful —bureaucracy! the Left!—to explain what it assumes to be insufficiently repressive force marshaled by police.

In the *Telegraph*'s discourse, evidence of poor police conduct is quickly glossed. On 13 August, the extent of the *Telegraph*'s reporting of this story is as follows:

> The police watchdog admitted it may have wrongly led journalists to believe that Mark Duggan, whose death triggered the riots in London, fired at officers before he was shot dead. The Independent Police

> Complaints Commission said it may have "inadvertently" given misleading information. (*Daily Telegraph* 2011b: 2)

Forty-three words constitute the *Telegraph*'s coverage—and even these fewer than four score words are qualified by hand-waving with two uses of the probabilistic "may have." By contrast, *The Guardian* devotes far more coverage to the circumstances surrounding Duggan's death that includes investigation into whether he was armed (Dodd 2011). Moreover, while Duggan's funeral registers a barely detectable 58 words in the *Telegraph* (*Daily Telegraph* 2011d), *The Guardian* devotes 800 words to the event, with extensive testimony from inside Tottenham (Muir & Taylor 2011). *The Telegraph*'s preferred terms of Us/Them dichotomizing are facilitated when the proximal trigger for the riots is obscured; namely, death at the State's hands.

In light of false police statements after wrongful instances of lethal force against citizens (Jean Charles de Menezes in 2005, Ian Tomlinson in 2009), the *Telegraph* could press harder for truth from the State. To do so would seem to be demanded by the ostensible "conservative" ideology of a restrained State as well as the political right's often-incanted valorization of law and order that should presumably extend to the agents of law-and-order. Instead, two pages after the 43 word item on Duggan having been fatally shot, the *Telegraph* devotes almost 900 words to US police chief and UK consultant Bill Bratton under the headline "Supercop's Battle Order." The article cites Bratton in extensive paraphrase while banging on traditionalist themes: "repercussions for those who step out of line must be severe," "police cannot be social workers," a bravely iconoclastic sentiment until one realizes that no one has proposed that they should be (Swaine 2011: 4). In sum, for *The Telegraph*, there is no confusing the virtuous Us with Them—and police are Us in a uniform.

Police on Their Backs: *The Guardian*

A number of reports in *The Guardian* are largely sourced via the police (*e.g.*, Laville, Dodd, Hawkes, Taylor & Walker 2011). The results of talking to police are, on occasion, surprising. While the *Telegraph* construes heightened confrontation with rioters as transparently commonsense policing, *The Guardian*'s Sandra Laville consults with president of the Association of Chief Police Officers Sir Hugh Orde. He is also former chief constable of the Police Service of Northern Ireland where plastic bullets and water cannon had been deployed. Contra the *Telegraph* editorial line, Orde comments,

> "I do not think it would be sensible in any way, shape or form to deploy water cannon or baton [plastic] rounds in London. (...) I would only deploy them in life-threatening situations. What is happening in London is not an insurgency that is going to topple the country." (quoted in Laville 2011a: 7)

This is one of several instances where Laville, *The Guardian*'s crime correspondent, reports views sourced by police that contradict the "tough" rhetoric that the *Telegraph* employs while presenting itself as speaking on behalf of police ("Our") interests.

The Guardian's position on the police—that they are needed and because of this must be held to a high standard—is expressed in Patrick Barkham's interview sketch with former Metropolitan Police commander Brian Paddick. In the course of 1,870 words, Paddick claims that, "'Even though I'm a liberal, the police should have gone in much harder'" at the onset of the riots. He adds that, "'These are people who if you say "boo" to them loudly enough, they will run away.'" Thus, Paddick advocated "'upping the ante to a level where they don't want to play any more'" in part via deployment of plastic bullets (Barkham 2011: 33). While reasonable people can disagree on this tactic, Barkham's report endows the use of plastic bullets with some logic for the purpose of maintaining order. However, and in significant contrast with the *Telegraph*'s premise of rioters as a hardcore of recondite Others, Paddick posits many rioters as hangers-on who would disperse if one says "'"boo" to them loudly enough.'"

The Guardian explores alternatives to the ostensibly macho, "shoot first, ask questions later" mode that the *Telegraph* endorses. Laville (2011b) presents a portrait of Karyn McCluskey whose concepts for countering gangs in Glasgow have been notably successful. In this model, police efforts are tightly focused for preventive purposes on recalcitrant cases while offering "carrots" for exit from gangs. Results for public safety have been positive as measured by reduced crime, in part because McCluskey's program plays on people's need to be in groups and channels them toward pro-social ones. The program and its concern with tangible results, as well as *The Guardian*'s attention to it, contrast starkly with the *Telegraph*'s demands for forceful suppression that is assumed to cow Them into submission.

Despite the *Telegraph*'s ostensibly robust support for police forces and for the Cameron government, the paper's posture appears to be chimerical. In particular, the Cameron-led government's drive for budget austerity implicates deep cuts in monies for police. *The Guardian*'s Watt, Wintour & Dodd (2011) report that a series of figures—from the Conservative Party London Mayor to the Vice Chairman of the Police Federation in England and Wales—express deep concern over the impact of cuts on police effectiveness. In the corpus of articles, the *Telegraph* appears nevertheless untroubled about cuts (one exception: Middleton 2011). Predictably, by early 2012, evidence had accumulated of a spike in crime since Cameron's premiership began doling out the empty bowl of austerity (Helm 2012). By contrast with the *Telegraph*, *The Guardian* not only discusses reduced monies to police services but the concurrent cuts to youth programs in vulnerable areas (Williams 2011).

The Guardian's coverage indicates that the *Telegraph* discourse missed opportunities to display contemporary conservatism's often proclaimed anti-Statism. Aitkenhead (2011) reports, for example, that a man looted an ice-cream cone, licked it, did not like the flavor, and gave it to someone else—and was subsequently jailed for 16 months for the offense. In this case, alleged law and order is a sadistic charade, more so as the contemporary right construes higher taxes on billionaires as a full blown Statist rampage. *The Guardian* also reports that many youth offenders were not only given stiff sentences, but that protocols for dealing with them (for example, specifically trained magistrates) were largely ignored. As Mark Johnson observes in *The Guardian*, "Young people who might have been helped to live differently are now in jails (...) to rub shoulders with career criminals and murderers" (2011: 37).

Riots, Communities and Victims Report

In reacting to parliament's interim report of Parliament's Riots, Communities and Victims Panel, *The Guardian* published two articles on 29 November. In the first article, Caroline Davies describes the report's conclusion that "The riots in August would not have spread from London to other areas in England had the initial police response been more robust in the capital" (2011: 11); this conclusion resonates with Paddick's analysis, discussed earlier. Davies characterizes the parliamentary panel's report as acknowledging that the police's (face saving, but wrong) statements about the circumstances of Duggan's death exacerbated negative public reaction. As with many newspapers, the establishment has its say in *The Guardian.*

In a different register on *The Guardian*'s "Education" pages, Chris Arnot presents an interview sketch with Professor Simon Hallsworth that addresses the same report. Hallsworth is described as "the UK's leading expert on gang culture" and his inclusion in the discourse illustrates *The Guardian*'s tendency to cast its net widely for sources. Hallsworth claims that Cameron's responses to the riots were "'scapegoating'" and stigmatized blacks and the "'feral underclass'" (Arnot 2011: 41). As to deep context, Hallsworth fingers "'the perverse form of capitalism'" located in neoliberal doctrine as having set the tempo of events, in part for making upward mobility a "'a thing of the past.'" Turning specifically to the report, the professor dismisses it as "'a superficial glossy brochure'" that channels the spirit of *Roshomon* seemingly in the hope that recording what enough people said will coalesce into analysis on its own. By contrast, Hallsworth locates the riots in the shift across generations from a "welfare state" committed to full employment to a "security state" that treats the citizenry's labor power as merely a cheap and flexible factor of production. Hallsworth's analysis is far removed from the personalizing and otherizing discourses of the *Telegraph.* As loathsome as rioting behavior was, *The Guardian* interprets it as a warning about the impact of neoliberal economics in the mass response to a darkening socio-economic climate.

Decoding Lammy

David Lammy, the Labour Party Member of Parliament from Tottenham, garnered coverage in both papers, albeit with markedly different imprimaturs. Lammy not only represents Tottenham, where the riots began, but also grew up there with the challenges of being young in a deprived area. As for the two the newspapers' postures toward Lammy, these reverberate with their editorial trajectories. Whereas Lammy is simplified or even instrumentalized into a traditionalist law-and-order advocate in the *Telegraph*—*i.e.*, Lammy as "one of Us"—more of the texture of his views are represented in *The Guardian*.

In a *Telegraph* editorial in November, Fraser Nelson (2011) rallies to Lammy's book on the riots as "fearless" and welcome "heresy against Labour's revered secularity" (2011: 32). Lammy is also discussed at length in Allison Pearson's *Telegraph* column that lionizes the MP as a child of Tottenham who was compelled to go to church each Sunday. Pearson instrumentalizes Lammy in order to polemicize about what she assumes to be Their convictions: "Unlike many in the Houses of Parliament, Lammy does not share a reflex contempt for Christian belief" since "it saved him from a life which he feared was destined to end in jail" (2011: 33). She also condemns "social liberalism and the rampant free market" for privileging materialism over spirituality. This gesture at least brings her into closer alignment with parts of Lammy's message and away from the *Telegraph*'s line by recognizing that the unbridled market is corrosive toward tradition. Nonetheless, as elaborated below, the *Telegraph* is selective in its uptake of Lammy.

As with many matters that surround the riots, Lammy is far more covered in *The Guardian* than in the *Telegraph* (e.g., Gentleman 2011). In this vein, Simon Hattenstone's sprawling and textured 2,460 word sketch of the Member of Parliament disrupts the *Telegraph*'s ventriloquism of parts of Lammy's message that it takes as serviceable in Us versus Them terms. The *Guardian*'s Hattenstone presents Lammy as intensely engaged as he interrupted his August vacation to return to his constituency *after* Duggan was shot and *before* riots erupted. Via Hattenstone, Lammy's account of the riots is highly critical of aspects of police performance. Lammy emphasizes that Duggan's relatives were not afforded the "'decency'" of being informed promptly and directly about his death (quoted in Hattenstone 2011: 6); instead, they learned about it through news reports. Beyond openly acknowledged mistrust of police from his youth, Lammy underscores that "'there is a history in Tottenham that involves deaths in police custody.'" Lammy also points to cross-service problems in policing. Whereas more responsive policing had been partially realized in the locale, Duggan was killed by Operation Trident (dealing with gun crime) run out of Scotland Yard. In Lammy's appraisal, the police "outsiders" to the neighborhood often exacerbate problems between Tottenham's public and the police. This level of detail on the policing back story never troubles the *Telegraph* for whom an undifferentiated police force constitute Our gallant centurions whose only failing is to be insufficiently coercive.

As coverage in *The Guardian* demonstrates, Lammy is an ideological weathervane of sorts. While he lambastes the Thatcher-Reagan regimes, Lammy appears to accept some neoliberal premises about implementing an agenda of "freedom" (Hattenstone 2011). Although presented as rhizomatic in *The Guardian*, Lammy does not much resemble the caricature of him as a one-dimensional evangelist in the *Telegraph*'s accounts.

Reading "Reading the Riots"

In the months after the brutal conflagrations, *The Guardian* inaugurated its "Reading the Riots" project. It is described as having been inspired by the media/academic collaborative investigation of Detroit's 1967 riots and, indirectly, by the Cameron government that "resisted calls for a public inquiry into the August riots" (Lewis 2011b: 11). The paper launched a "grand opening" of the project with five articles on Monday, 5 December that totaled more than 7,600 words. This level of investigative coverage continued through the week. As of May 2012, Reading the Riots is an extensive and dedicated feature within *The Guardian Unlimited* webpage.

Given its commitment to a simple and largely monocausal narrative about the civil disturbances, *The Daily Telegraph* angers over Reading the Riots. Philip Johnston's apoplectic response features a brigade of Us/Them terms, with references to "scum of the Earth," "a rag-tag army of opportunists, feral criminals," and "ne'er do wells" (2011b: 29). However, his main editorial target is *The Guardian* that he takes as aligned with Them. While "the rest of us cheered to the rafters" over harsh sentences, he proposes that "the Left seethed with indignation" (2011b: 29). Johnston is livid over rioters' criticisms of Our avatars, the police: "One rioter told the researchers, presumably while struggling to keep a straight face: 'The police is the biggest gang out here.' *If only*" (emphasis added, 2011b: 29). In this moment, Johnston openly thrills to the thought of the State meting out unbridled mayhem against Them, conservative hymns to law and order notwithstanding.

While the *Guardian's* Reading the Riots project at times shades toward a *Roshomon*-like cacophony, it perturbs easy conclusions.[2] Topping & Bawdon are incredulous toward some informants' accounts of the "high minded" causes for which they rioted. At the same time, they report rioters' regrets that an opportunity for a more constructive confrontation had been squandered (2011). The series of articles also presented brief portraits of victims of the rioting: a 35-year-old woman whose house was burned down (Adegoke 2011), a man whose shop was decimated (Ferguson 2011). Figures reported in the series of articles fill in further back story. Fifty-nine percent of rioters were unemployed (Lewis, Newburn, Taylor & Ball 2011) and, according to the government's own figures, two-thirds had special educational needs (Travis 2011). *The Guardian*'s reporting also unearths figures that are congenial, when taken in isolation, to the right-wing's one-note narrative of antisocial people at their worst. Twenty-five percent of rioters are reported as having been previously convicted of ten or more offenses and 77 percent of adult rioters had prior criminal records (Travis, Ball & Bawden 2011). Not surprisingly, the disturbances generated an enticing recruiting call for incorrigible criminals. Alongside hardened elements, more than half of the rioters were under 20 years old and 21 percent were reported as being younger than 17.

Under the Reading the Riots rubric, Raekha Prasad (2011) reports on the depth of ill-feeling toward the police. Among the stories presented are routinized police incivility, "stitch ups" for crimes the accused did not commit, and accusations of harassment toward people as young as 12 (which, if true, set up a lifetime of antipathy toward police). Seventy three percent of *Guardian* informants claimed to have been subject to stop and search in the previous 12 months. Elsewhere, *The Guardian* reports more than one million police "stop and searches" in 2008-09 with strong racial asymmetries in their applications in some districts (Lewis & Dodd 2011). In this view, the rioters are first otherized by policing practices while right-wing discourse, as in the *Telegraph*, swoops in to otherize all over again. In rejecting otherization, Reading the Riots spotlights dysfunctions that a cohesive society must address with more imagination than is presented by sanctimonious editorial tirades and draconian jail sentences.

The Guardian's Lewis & Dodd seek police response to Reading the Riots and encounter their "'willingness to learn'" better practice (Lewis & Dodd 2011: 8). In this manner, *The Guardian* assays improved calibration between public safety and civil liberties that contrast with the *Telegraph*'s caricatures of police (White Knights) and elements of the public (incorrigibly feral underclass). The project is of a piece with *The Guardian*'s refusal of the Us/Them binary. It is, moreover, significant if journalism is to have a part in the construction of an improved social contract, rather than participating in a cycle of blaming Them and papering over deep-seated, extra-personal social fissures.

Conclusion

This paired case study of *The Guardian* and the *Telegraph* illuminates several matters of interest to the Propaganda Model. First, Herman & Chomsky's original anti-communism filter was highly salient in the US during the "neo-Cold War" of the Reagan era 1980s. Retrofitting the now largely archaic anti-communism filter to a more abstracted Us/Them dichotomy extends the model's reach across myriad socio-political issues. The Us/Them dichotomy may, furthermore, be exported to an analysis of any society's news media while also noting They are not necessarily people from beyond a nation's frontiers.

Second, the paired case study of *The Guardian* and *The Daily Telegraph* suggests (even if it does not prove) that ownership patterns and the demands of commercialism condition content. While the *Guardian* is a paid daily newspaper, and actually contains more pages devoted to advertising than the *Telegraph*, it is in some degree buffered from market-driven practices through subsidy from the Scott Trust. In making the comparison between the partly subsidized *Guardian*, that need not strictly follow market dictates, and the *Telegraph*, substantial differences in news coverage are evident. These differences extend to coverage of the broad ambiance of the political-economic milieu. To wit, while the *Telegraph* finds only individual malfeasance, *The Guardian* is attentive to neoliberal pressure on people who must live through market relations as an aggravating factor in the riots.

Third, in making the comparisons in detail between coverage of a multifaceted issue in *The Guardian* and *The Daily Telegraph*, it is apparent that the left-leaning paper's coverage is not merely more voluminous. It presents superior journalistic performance with respect to informing the audience on an array of riots-salient topics. In this view, the ideology of the paper also matters and, at present, all ideologies are not equally ideological or acute in their purchase on the truth. During a generation in which it has contemptuously attacked the "reality based community" (Suskind 2004), adopted the most anti-intellectual forms of post-modernism (Wolfe 2000), and attempted to corrupt even scientific methodologies (Mann 2012; Oreskes & Conway 2010), the right has seceded from the Enlightenment environment in which truth is known through empirical inquiry. In this view, it follows that the news organs that hail the right will exhibit signature failures in locating the contours of reality; a theme that will be reprised in the next chapter. Fourth and finally, in pulling together Chapters 4 and 5, effectively right-leaning discourses (*The New York Times* on Iraq, *The Daily Telegraph* on the riots) exhibit striking continuities in the persistence of their Us/Them rigidities and their investment in the exercise of State force as solution.

Notes

[1] Letters to the Editor" also produce striking contrasts between *The Guardian* and *The Daily Telegraph*. As right wing as the *Telegraph*'s editorial line is, the letters section often exceeds it. To take one example: Shooting looters on sight is advocated by one *Telegraph* letter writer, citing alleged Jamaican methods of law

enforcement as a model to emulate (Schofield 2011). In stark contrast, many *Guardian* letter writers point to double-standards surrounding the riots and the subsequent regime of punishment. As concerns Cameron's open lenience toward an advisor with serious legal problems, before and after his appointment, one letter writer observes, "It is not hard to see this as one ethos for the rich and powerful and another for the marginalized" (Beattie *et al.* 2011: 35).

[2] While Reading the Riots generally complicates the Us and Them dichotomy, and shifts the emphases away from individual monads to a highly textured context, in some moments it also simplifies by striating the population into various "interest groups" (along the lines of religion [Malik 2011] and race [Muir & Adegoke 2011]). In what reads as a contrived form of recognition, an article on gender and the riots observes that, while 90 percent of the participants were male, "girls and women (...) appear to have played a significant role" (Topping, Diski & Clifton 2011: 16). Nonetheless, as these categories of identity are already given in contemporary discourse (religion, race, gender), it is not surprising that *The Guardian* reaches for them in its investigative/explanatory mode.

CHAPTER 6
To "Tell the Truth!" in Flak Style

The world changed palpably for the better on 10 October, 1998, when Spanish magistrate Baltasar Garzón Real issued a warrant for the arrest of Augusto Pinochet while Chile's retired dictator was in London. Although the UK refused to remand Pinochet, Garzón's bid to apply universal jurisdiction reverberated. In Latin America, notably, Garzon's action prompted the collapse of the impunity in which abusive regimes had been formerly ensconced. Beyond the Pinochet warrant, Garzón's investigations have fingered the illicit drug industry, Silvio Berlusconi's media empire, Spanish government corruption such as the Gürtel affair, the violent Basque separatist group *Euskadi Ta Askatasuna* (ETA) as well as the *Grupos Antiterroristas de Liberación* (GAL)'s extrajudicial actions against ETA. The biggest prosecution yet undertaken against al Qaeda also emerged from Garzón's office and netted 18 convictions (Cox 2012; Turley 2011).

In light of a sparkling career in which the judge employed the lawful arm of the State to advance human rights, it may be a surprise to find that at the relatively young age of 56, Garzón is today involuntarily retired. How and why has this happened? During the Transition that followed Francisco Franco's death in 1975, Spain implemented a "pact of forgetting" to facilitate going forward as a nation. The pact was given the force of law and immunized Franco-era figures from prosecution. Since the turn of the millennium, however, grassroots efforts have increasingly questioned the terms of forgetting. In turn, Garzón's move to confront the past and prosecute Franco-era crimes triggered lawsuits against him from extreme right-wing organizations (*Manos Limpios* and *Falange Española*) that led eventually to his disbarment.

Reed Brody of Human Rights Watch comments that, "'Thirty-six years after Franco's death, Spain is finally prosecuting someone in connection with the crimes of his dictatorship—the judge who sought to investigate those crimes'" (Human Rights Watch 2012: 1). Moreover, prosecution of judges is "almost unheard of in Spain" since their independence is an important guarantor of rigorous advocacy against a society's power center institutions (Human Rights Watch 2012: 2). Garzón's case implicates the tactics and strategies of harassment that Herman & Chomsky discuss under the rubric of the flak filter. More specifically, Herman & Chomsky characterize flak as "'negative responses to media statement or program" that is enacted to shape and discipline media behavior (1988: 16). Phone calls, letters, editorials, and faxes directed toward a target or his or her employer may all do work of flak—as well as tweets, emails, and web pages devoted to ideological takedowns. Flak may become more serious still when it crystallizes into investigations, lawsuits, or regulatory action.

In considering Herman & Chomsky in the present, one may expand their concept beyond flak that is directed against media organizations. At the same time, all flak campaigns necessarily present a media dimension, whether or not the resultant flak is finally directed at a media organization; that is, flak against a political movement or an outspoken artist necessarily draws on media. As a species of highly charged criticism, up to hair-on-fire polemic, flak may also traffic in emotionalized claims that have an agitational, action-oriented impact on audiences.

In further exploring the concept, it may be reasonably asked whether flak can be adopted as a tool of the weak to wage asymmetric struggles. In this view, flak can be construed to include demonstrations, boycotts or letters to the editor when, for example, readers are bothered by chauvinistic discourses in a publication. Notice, however, that Herman & Chomsky posit flak as "related to" and imbricated with "power" (1988: 17). In order to preserve the negative connotation of flak's alignment with authority, one needs to differentiate flak from activism against prevailing conditions. While their tactics may be similar, structural considerations and strategic objectives are the grounds on which to distinguish flak from activism. In this view, activism may be defined as actions performed in good faith toward pro-social ends, typically in confrontation with authority. Flak, by contrast, trades on bad faith, is malicious as needed, and is often utilized to thwart pro-social change in favor of the status quo or for reactionary purposes. In this view, flak often carries water for authority or factions of it. Referring to the earlier example, Garzón's legal quests for human rights may be construed as a species of activism, whereas prosecution of him on highly contentious grounds constitutes flak. It bears further notice that, just as grassroots activism can be simulated in what have been called "astroturf" campaigns (see Greider 1992: 36-39), flak on behalf of the powerful can be readily marshaled and endowed with a patina of populism.

In this volume, I have suggested that some filters are more powerfully in play at present than they were when Herman & Chomsky introduced the Propaganda Model. Ownership is more concentrated than in 1988, as Noam's (2009) empirical treatment attests. Moreover, evidence strongly supports the claim that reporters are currently more dependent upon information subsidy from elites and the PR industry due to the expansion of the news hole (in part, through technologically enabled platforms). On the other hand, the anti-communism filter no longer operates as it once did, even as Us/Them dichotomization continues to perform yeoman work in structuring news narratives. While difficult to quantify, flak's importance and prevalence appear to have intensified in the years since the original formulation of the model. If nothing else, the channels into which flak may be inserted have increased and its cascades across proliferating media channels intensified. The production of flak is not, moreover, bound by the conventions of journalism that at least gesture toward procedural pursuit of objectivity and fairness (even if these tend to rally to the status quo).

In this chapter, I will examine the production of flak via an extended case study of an organization—Media Research Center—that is dedicated to placing flak into the media pipeline. I drill down into the 18 most recent "Special Reports" posted on the organization's web page as of 12 June 2012. As the Media Research Center (MRC) web page touts its reports as "in-depth study, analysis or review exploring the media," one can assume that they are supposed to constitute carefully constructed and sophisticated discourses. The 18 MRC reports sum to over 102,000 words, for an average of slightly less than 5,700 words per report. In assessing the reports' methods of inquiry and measuring their claims against other sources to evaluate truth value, I argue that MRC's discourses generate flak and not insight. In turn, I posit that MRC's flak performances are made to order for insertion into other news discourses for the purpose of buttressing right-wing causes and doctrines. These include the right's party lines that posit media as a monolithically left-leaning bloc, climate change as a fraud, and George Soros' left-leaning philanthropies as a cover for pan-global domination. To further characterize the multifaceted dimensions of flak in preface to the extended case study of MRC, I open with a brief portrait of a media-enabled flak campaign against an academic researcher in the currently salient domain of climate change science.

Flak Case Study: Hot Air

How does a flak campaign play out in practice, from generating newsworthy events to stimulating grassroots participation? Consider Michael E. Mann, a climate investigator at Pennsylvania State University and one of the investigators who published the original "hockey stick" graph that exhibits the spike in temperatures at the end of the twentieth century. In 2005, Congressman Joe Barton (Republican-Texas) initiated a flak campaign against Mann via extensive requests for research materials on a presumption of misconduct. Barton's move necessitated "legal advice and representation" on Mann's part (Mann 2012: 151). The demands from a seat of political power conveyed a message of intimidation with fist-in-the-face clarity that also turned Mann away from the conduct of his work and to the defense of it. As part of the campaign, Barton commissioned academic statistician Edward Wegman to compose a report to debunk Mann. In short order, Wegman disgraced himself with a conviction for plagiarism (*Deep Climate* 2011; *Nature* 2011).

In 2009, emails that included Mann's correspondence were stolen from a back-up server at the UK's University of East Anglia and published on the internet. As Mann's employer observes, "Beginning on or about November 22, 2009, The Pennsylvania State

University began to receive numerous communications (emails, phone calls, and letters) accusing Dr. Michael E. Mann of having engaged in acts that included manipulating data, destroying records, and colluding to hamper the progress of scientific discourse around the issue of anthropogenic global warming" (Foley, Scaroni & Yekel 2010). Along with threats of investigations and demands to submit materials, Senator James Inhofe (Republican-Oklahoma) released a lengthy report to the Senate web page calling for criminal prosecution of several climate scientists including Mann. Republican Congressmen Darrell Issa (Republican-California) and James Sensenbrenner (Republican-Wisconsin) pressed for Mann to be investigated for the purpose of slashing his National Science Foundation funds (that were, in turn, speciously conflated with Obama stimulus programs, as the Republicans flacked at two birds with one stone).

Outside officialdom, the flak response was bare knuckled. Mann reports having received an envelope with white powder (2012: 227-228). While the material turned out to be innocuous (corn starch), the resultant anthrax terror scare was not. A former CIA agent solicited other Penn State professors with reward money to "whistle blow" on Mann, on the presumption that there was something to blow the whistle about. Flak was also exerted at the state level as the Commonwealth Foundation lobbied for Penn State to have its funding cut by the legislature until the university fired Mann. The same foundation ran a week of attacks on Mann in the university's student newspaper. At one of Mann's presentations at a conference in Pittsburgh, he was subjected to "a billboard-sized display bearing an unflattering cartoon" as well as tee-shirts for sale that mocked him (Mann 2012: 232). Meanwhile, as University of East Anglia (2010) observes, a series of at least five investigations (The Commons Science and Technology Committee, Oxburgh Panel, Russell Panel, Environmental Protection Agency, Penn State) on two continents had exonerated the scientific endeavors of the researchers whose emails had been stolen. Perhaps the most appalling aspect of the multidimensional flak campaign against Mann, in which operators in the tree-tops stimulated the grass roots, was that the professor and his colleagues were, by all rigorous and fair-minded accounts, scrupulously conducting their scientific dialogues with nature.[1]

Having considered the impact that flak may have on its targets, and how little grounding in evidence that it may possess despite the strength of its claims, I now pivot to examine a dedicated flak producer.

Media Research Center: A Flak Production Line

In introducing itself on its webpage, the Media Research Center (MRC) places its flak mission front and center. The organization self-describes as "proud to celebrate 25 years of holding the liberal media accountable for shamelessly advancing a left-wing agenda, distorting the truth, and vilifying the conservative movement" (Media Research Center n.d.: 1). Despite presenting itself as on a lonely and even defensive mission within a hostile media ecology, MRC reports 63 full time staff (2008: i), revenue of almost $(US)11 million and assets that value at close to $(US)10 million, all of which speak to a decently endowed organization capable of going on offense (Media Research Center 2008: 29). Moreover, although it also presents a deeply troubled relationship toward truth and scholarly rigor, MRC's exhortation is repeated at the end of each of its reports: "*Tell the Truth*!" (original emphasis, Media Research Center n.d.: 1).

MRC adopts many of the conventions of think tank and academic reports such as "Executive Summaries" with similarly aggrandizing titles for staff such as "Director of Media Analysis." Furthermore, like an academic college, the organization is parsed into departments. These include the News Analysis Division, Business and Media Institute, Culture and Media Institute, NewsBusters, and TimesWatch. Notwithstanding its impersonations of academe, MRC departs from academic practice for the straightforward reason that its studies are designed for purposes other than contributions to knowledge. MRC's special reports present, from conceptualization to the dissemination of results, as vehicles for flak. Moreover, the reports exhibit no signs of quality control through the external review that is an important feature of the vetting process for academic literature. Internal vetting also appears to be minimal at best. For example, Rich Noyes and Geoffrey Dickens' (2011) report on morning television programs ostensibly covers 723 stories aired from 1 January through 31 October 2011. Noyes & Dickens' report was posted on MRC's webpage by 15 November. In other words, the report on over 700 stories gestated for a mere two weeks after all the data was in before being published on the organization's web page.

The MRC's "About Us" section of its webpage does not disguise its project of one-sided indictment. Part of the self-characterization reads:

> [The] News Analysis Division [is] the "Leader in Documenting, Exposing, and Neutralizing Liberal Media Bias." (...) [It] published the daily CyberAlert e-mail report, weekly Media Reality Check reports and the every other week Notable Quotables collection of the most biased quotes from journalists—including the annual year-end "Awards for the Year's Worst Reporting"—as well as "Profiles in Bias." (...) Also part of the division: The MRC's annual "DisHonors Awards" gala (...). (Media Research Center n.d.: 1)

Devotion to the project of flak is underscored in the 2008 *Annual Report* when MRC promotes a "gotcha" story among the year's triumphs. One of its "trophies" is to have reported that "House Speaker Nancy Pelosi (D-California) had been using a bogus bible quote to promote global warming legislation." This presents as an explosive story in MRC's estimation, although its "gotcha" exhibits none of the tenacity and skill demanded of genuine investigation (Media Research Center 2008: 12).

Substance aside, how successful are MRC's efforts as measured by reaching the media pipeline? While one should regard figures reported by MRC with caution, the organization's founder and President L. Brent Bozell III claimed in 2008 that "MRC material was cited by blogs, newspapers, news services, TV news, magazines and other news sources 9,473 times, an average of 31 times a day" (Media Research Center 2008: 1). Bozell's arithmetic notwithstanding, 9,473 citations divided by 366 days in leap year 2008 actually comes to 25.9 cites per day. Although Bozell derides what he calls the "liberal media-government axis" (Media Research Center 2008: 1), the *Annual Report* boasts that "experts" from its News Analysis Division logged appearances on 86 television and 682 radio programs (Media Research Center 2008: 3). By MRC's own account, the media that mainly absorb the organization's products are avowed right-wing channels such as *New York Post*, *O'Reilly Factor*, *Hannity & Colmes*, *Your World with Neil Cavuto*, *Fox and Friends* (all Murdoch/News Corporation owned), *Washington Times*, *Human Events*, *Townhall*, *Newsmax*, *WorldNetDaily*, and *The Rush Limbaugh Show*. Despite the stated hostility toward mainstream media as part of an "axis," leading exemplars of it such as *Washington Post*'s Howard Kurtz (2004) also endow MRC with respectful citation and do not cross-examine the organization's goals or methods. Wherever it circulates, MRC appears to be a quasi-academic enterprise, given its name and regardless of the quality of its products or organizational mission.

General Tendencies in MRC's Flak Discourses

Along with efforts to contrive flak under its "*Tell the Truth!*" tagline, MRC often mints crass discourses that include tasteless jabs at gay people (Graham & Dickens 2011: 2; Philbin 2012: 4) and dog-whistled reference to Obama's early years in Islamic Indonesia (Media Research Center / Culture & Media Institute 2012: 2). Right-wing pillars of traditional authority are, at the same time, beyond any shade of criticism, as when MRC dismisses questions about the Catholic Church hierarchy's complicitous response to pedophilia in its ranks and its retrograde posture toward women (Graham 2012: 10-11).

Nonetheless, does MRC get anything right? One kernel of truth in their reports is the observation that Obama's foreign policy has prompted relatively muted criticisms in news media. Some legally sophisticated bloggers (for example, *Harper*'s Scott Horton 2012) and international news coverage by summer 2012 (Harris 2012; Milne 2012) address the extent to which Obama has moved assertively into the legal and ethical twilight of lethal drone missions and stepped-up State secrecy. However, by contrast with Bushian neoconservatives' penchant for aggressive chest-beating about its "tough guy" postures, Obama's team has managed the climate of opinion via lower-key tactics. Mainstream news media has, as a result, been far too complacent toward the Obama administration's conduct of foreign affairs and channeled a "sphere of consensus" that normalizes state secrecy and permanent, "low-intensity" US interventionism. MRC takes these issues up, however, only to score cheap debating points about a brand of Statist policies to which it otherwise gives zealous support. MRC's gestures are substantively empty and driven by the calculus of supporting or attacking US administrations based on their party affiliation and not on their alignment of deeds and principles. As such, MRC's disingenuous arm-waving contrasts with voices that demand human rights as a matter of baseline commitment (for example, the previously discussed Garzón, who forthrightly criticized US adherence to international law in both the Bush and Obama eras [Turley 2011]).

Heads, I Win, and Tails, You Lose: Double Standards

Although double standards present one of its central indictments toward mainstream media, MRC exhibits an audacious tendency to fashion or apologize for them as doctrine demands. In doing so, MRC repeatedly claims that a double standard is evident when avowedly conservative figures are described as "conservatives" and/or via their religious views (Noyes & Dickens 2011; Media Research Center / Culture and Media Institute Staff 2012). For example, MRC reprimands a mainstream media reporter who characterized Sarah Palin and Michelle Bachmann as "'the queens of the Tea party. Tough, uncompromising, as conservative as they come'" (Noyes & Dickens 2011: "'Conservative' Republican vs. Non-Ideological Obama?": 2). As figures such as Palin and Bachmann have made strenuous efforts to portray themselves as stridently to the right, it is difficult to find a warrant for complaint with characterizations that echo and even valorize them for it.

MRC also does not ask, thus does not answer, whether Republican candidates invite attention to religion by constant and ostentatious references to their religiosity. In this view, MRC does not report research findings as much as demand that Republican candidates be allowed to have it both ways. To wit, Republicans must be ever moving further rightward, ever more aggressively in tension with the Establishment Clause's firewall between Church and State. And, at the same time, MRC prescribes that Republicans must be presented as plain speakers of common sense, wholly innocent of ideology. Noyes & Dickens indignation also rings hollow when they castigate the networks for having even indirectly dwelled on “fringe behavior” from Republicans when the party invites such characterizations by harboring viscously stupid, flak-driven discourses about, for example, the validity of Obama’s US citizenship (2011: “Interviews”: 5).

Despite its complaints, even obvious double-standards do not trouble MRC if they advance the right’s game of “heads, I win, tails, you lose.” Wealth and economic deregulation are valorized, as befits MRC’s economically neoliberal party line. However, via the alchemy of double standard, MRC derides the capitalist enterprises of Hollywood for the riches it has realized via the profitable trafficking in vulgarity that capitalism generally unleashes (Philbin 2012).

Along with double standards about the market, MRC also constructs them with respect to the State. In this vein, MRC’s Tim Graham revives dreary talking points from the early Bush era in support, for example, of the facility at Guantánamo Bay (2011: 4-7). Despite tough rhetorical stances assumed by right-wing punditry, the most searing criticism of Guantánamo’s brand of Statist abuse arises from establishment realms with intimate knowledge of it; most notably, the military (Taguba 2008) and figures tasked with implementing procedures at Guantánamo such as Chief Prosecutor Colonel Morris Davis (*The Moderate Voice* 2012) and prosecutor Lieutenant Colonel Darrell Vandeveld (Sullivan 2008). Colonel Lawrence Wilkerson (2009), chief of Staff to Secretary of State Colin Powell, compiled a particularly livid inventory of the Bush administration’s reckless amateurism at Guantánamo. Risible conduct included failure to focus on actual terror suspects by incarcerating innocent bystanders tragically caught up in the dragnet alongside hardened operators. An insider to the administration, Wilkerson also claims that Bush’s inner circle knowingly lied to the public about the useless intelligence value of almost all of Guantánamo’s prisoners in order to efface its own record of security failure. Despite the dishonor and the thundering ineptitude associated with Guantánamo in the eyes of principled establishment figures, MRC continues to flog at the issue as if it imagines itself bullying a shambling pamphleteer off a street-corner. Moreover, in this and other cases in MRC’s right-wing discourse, the State is urged to coercive action.

On the winds of double standard, MRC weathervanes from endorsing crude Statist abuse to an allegedly anti-Statist, neoliberal posture. While detaining suspects who have not been indicted of a recognizable crime for years passes muster as legitimate State activity in its estimation, MRC simultaneously maintains that "public broadcasting marinates in a very strong Statist ideology" apparently as an efflux of its mere existence (Graham & Dickens 2011: 20). In all of this, MRC's double-standards as concern the State align with orthodox right-wing positions.[2]

A Madness to Their Method

Despite the fact that its name is Media *Research* Center, the organization exhibits dire problems in producing material that is recognizable as scholarly investigation. Expanding on Media Matters' (2005) pertinent criticisms, MRC's research findings are as predictable as "elections" in an authoritarian one-party state. A basic problem with MRC's reports is that they offer no structural model to explain media behavior or to constrain and order assumptions about it. This is so because, to all appearances, the reports that MRC publishes are not designed to present contributions to knowledge but are generated for the purpose of ideological and partisan combat.

In lieu of a model, MRC unconditionally asserts that the US' private-sector-dominated news media is a permanent left-wing campaign and an adjunct to the Democratic Party. In this view, ownership patterns, private sector demands, professional mores, audience preferences, and the pull of events themselves are mooted; reporters simply cram left-wing dogma into every crevice of the news hole, shamelessly, without inhibition or brakes. For MRC, "activist reporters" are the norm (Waters 2010: "Introduction": 1); that is, when news media does not exhibit monolithic docility. In this vein, Julia A. Seymour writes, "Just as the Pied Piper's song entranced the rats, Obama's tax cut promises captured the attention of the news" (2010: 2). When they are neither activists nor rodents, news workers report for duty as "Obama's media palace guards" in Matt Philbin's indignant appraisal (2012: 2). For MRC, it is given that private sector media are left-leaning and that they have monolithically enlisted in Obama's campaigns, unencumbered by institutional or professional constraint.

MRC predictably scorns public broadcast as "a liberal playground" in which "seeking balance is strictly forbidden" (Graham & Dickens 2011: ii). However, MRC's fire is mainly trained on the dominant strain of private sector media in the US even as its discourses otherwise exalt the private sector. In an exceptional moment when the structure of media industries arises briefly, MRC adopts a populist idiom and indicts Viacom for taking "corporate synergy to the next level" (Graham 2010: 17). MRC infers synergy at work when Steven Colbert of Comedy Central's *Colbert Report* is interviewed on CBS's *60 Minutes* (Viacom properties all). However, in contrast with Herman & Chomsky, MRC does not take political economy seriously by situating it within an elaborated model. Indeed, synergy of the sort that MRC supposes in the Colbert case is a straightforward manifestation of the corporate dominated and deregulated order to which the organization otherwise devotes full-throated support. In this view, when political economy arises in MRC's discourse, it is for the purpose of scoring a cheap rhetorical point and not out of intellectually grounded conviction.

Other basic problems bedevil MRC products. MRC mimics academic research in minting positivistic claims such as, "The networks were critical of" Republican presidential candidate Rick Perry's "religious views 63 percent of the time" (Media Research Center / Culture & Media Institute Staff 2012: 3). However, the report is effectively silent on the substance of the methods by which the determination was made. To wit, the "Methodology" section is three sentences long and only addresses selection of transcripts—and *not* the procedures of MRC's content analysis that is effectively unreplicable despite ostentatiously enacted pantomimes of rigor (2012: 5). Moreover, as is expected of a content analysis, no reference is made to inter-coder reliability. Despite the fog of vagueness around its methods, MRC's conclusions are neon clear in asserting "an ugly exhibition of blatant journalistic bias and the secular elite's disdain for people of faith" (Media Research Center / Culture & Media Institute Staff 2012: 5).

Quote, Unquote

Further dire problems characterize MRC's reports. As concerns reporting results, MRC eschews complete citations for quotations or other information in a sources section. At times, there is no citation of any sort, as if pertinent information floats innocently through the air before becoming lodged in MRC's texts. For instance, Seymour (2010: 1) asserts that Obama tax increases will amount to $(US)4.2 trillion without any backing or citation—and although the claim that the administration has in fact raised taxes for the vast majority of citizens has been appraised to be false (Kessler 2012). MRC also regularly summons sources that are as committed to doctrinaire right-wing discourse as itself. MRC's regularly consulted sources include *The Washington Times*, *Big Hollywood*, Heritage Foundation and Fox News. On the rare occasions when MRC reports venture into academic literature, they typically do so via second-hand accounts of it in other publications. Hence, Seymour cites a *Forbes* column that quotes a scientist's discussion of toxins—but does not cite the original investigation (2012a). In eschewing examination of the original research, MRC's staff depends on "fellow travelers" and/or non-academic media sources in order to distill the original investigators' conclusions. In other words, MRC draws its information from the same media that the organization ostensibly audits, at the expense of parsing original sources.

During the sampled time period, MRC constructed several reports by marshaling cherry-picked quotations *en masse*. In their contribution to this MRC genre, Tim Graham & Geoffrey Dickens (2011) assemble the *20 Most Memorable Leftist Excesses of Public Broadcasting*. The report consists of 20 quotations culled from the massive inventory of discourse over public broadcasting across decades. A similar procedure—indeed, some of the same quotations—informs Graham's *Secular Snobs* (2012) published six months later. In each report, the technique is pure "gotcha" in shaking the prospector's pan in the archives for brief snippets of discourse. The technique does not differentiate between some genuinely ill-considered comments and ones that have been cleansed of back story; all are presented as self-evidently wrong by MRC.

Perhaps the most striking of the reports constructed from heavily vetted quotations is devoted to Ronald Reagan. As the last Republican Party president to leave office with popular support largely intact, Reagan has assumed the quality of "Maximum Leader" for the contemporary right, hence MRC issues a report in his "defense" almost thirty years after his final electoral campaign. In a strange research design, the *Rewriting Ronald Reagan* report itself re-writes writing on Reagan via 8,700 words *exclusively devoted to negative* quotations about him from news workers and celebrities. In what reads as a frantic attempt to contrive a martyr by inventing a media environment of monolithic hostility, MRC maintains that "the elite scorned" Reagan (Baker, Graham & Noyes 2011: Reagan the Man: 1). Nevertheless, the former Hollywood star, two-term governor of California and two-term US president was far more deeply ensconced within the elite than most anyone who lived during the twentieth century. Furthermore, assertions about media derision were decisively not shared by Reagan's media team that was pleased with the coverage that they had painstakingly staged (Hertsgaard 1988: 4). Unsurprisingly, MRC's appraisal of coverage of Reagan is also not corroborated by rigorous academic research (Entman 1989: 30-36, 46-74).

Case Study of MRC's Methods: Obama's Morning Emissaries

A more extensive case study of one report will serve the purposes of telescoping in on the methods that inform MRC's reports and that prioritize production of flak over interrogating reality. In the 8,000-word *Still Thrilled by Obama*, Rich Noyes & Geoffrey Dickens state that "analysts examined all 723 campaign segments" aired on the three main networks' morning programs across the first 10 months of 2011 (Noyes & Dickens 2011: "Executive Summary": 1). In a typical move, MRC excludes examination of News Corporation (Murdoch)'s Fox without further explanation. Noyes & Dickens write that, "In spite of the terrible economic situation, Barack Obama was still treated mostly as a celebrity, with the networks providing the President and his political team a forum to trash their competitors" (2011: "Introduction": 1). Beyond an apparent kernel of truth as concerns US media's penchant for soft-focus personalization (Bennett 2001), Noyes & Dickens assert that the *raison d'être* of the morning programs is to support the Democratic Party president with corresponding disparagement of Republicans. At the same time, they do not present compelling evidence that this has occurred, nor do they marshal any structural reason why it should be expected to occur.

Noyes & Dickens claim that 82 percent of the questions posed to candidates "reflected a liberal policy agenda" (2011: "Interviews": 2). Despite the apparent quantitative hardness of the datum, MRC furnishes no methods section to account for how the figure was generated, nor do they report data on inter-coder agreement. Moreover, all questions (100%) were coded as either liberal (82%) or conservative (18%). This scheme reads on its face as a simplistic dualism that fails to account for questions from the political center, that may be double-coded, or that are in a "devil's advocate" register. Indeed, the examples in Noyes & Dickens' text appear to be questions that test the consistency of politicians' positions and the depth of their convictions, rather than relentlessly interpolating them to left-leaning positions. While Noyes & Dickens catalogue questions, they are also unconcerned with the politicians' answers, as if the question is assumed to be the absolute determinant of the response. Despite striking methodological weakness, Noyes & Dickens (2011) manufacture flak toward network morning programs, Democrats and Obama that enables MRC to deliver seemingly rigorously fashioned numbers into the media pipeline.

Climate of Ignorance

I now turn to coverage of issues in MRC's reports as these intersect with frequent targets of flak for the right in general and MRC in particular. The two issues that I will take up are climate change and billionaire philanthropist George Soros, both subjects of multiple MRC reports. As concerns climate change, it has become a key arena of right-wing flak (Goss 2009: 464-65; Oreskes & Conway 2010: 169-215). Two MRC reports participate in the trend, even as they exhibit near complete silence on salient features of climate science. Instead, MRC cherry-picks general interest news reports on climate dating as far back as a century before any of the Intergovernmental Panel on Climate Change (IPCC) reports (Gainor 2010) and fixates on the person of Al Gore (Seymour 2011).

The center of gravity in Dan Gainor's report of more than 6,000 words is that news media cannot make up its mind about climate change. His method consists of rummaging through twentieth-century journalism for cherry-picked material in order to posit that news media has made mutually exclusive claims; notably, about whether cooling or warming will be the wave of the future. Gainor's "*New York Times*-line" features one-sentence quotations on climate from the newspaper dated 1924, 1933, 1975 and 2005 (2010: 2) while on the following page he presents more one-sentence snippets from *Time* magazine dated 1923, 1939, 1974 and 2001. Putting aside the obvious fact that reporters are not climate scientists, it was only in 1996 that the Intergovernmental Panel on Climate Change cautiously fashioned the claim of "'discernible'" influence of human activity on climate, above and beyond natural variation (Oreskes & Conway 2010: 204-205). These conclusions were further firmed up around evidence reviewed by the third and fourth IPCC reports in 2001 and 2007. By the 1990s, new lines of evidence and methodologies that did not previously exist were being brought to bear on the long sweep of climate behavior. For these reasons, it is untenable to posit, as Gainor does, that claims from news reports from the past are "Just like today" for what he asserts to be their arm-waving alarmism (2010: 6).[3]

While Gainor resolutely avoids encounters with scientific discourse generated by peer-reviewed reports, he does cite post-punk legends The Clash ("The ice age is coming" lyric from the "London Calling" single) and science fiction novels. As for action, Gainor demands that journalists follow the professional dictum to "Distinguish between advocacy and news reporting" and thereby asserts without evidence that the guidance had been widely flouted. Gainor also demands that coverage of climate change obligatorily include calculations of costs to ameliorate it, with the strange implicit premise that climate science must be declared "false" if being true costs too much. Furthermore, the UK's Stern Review makes such economic calculations about climate change—albeit, with emphasis on the staggering opportunity costs of *not* blunting climate change while it is still possible. The Stern Report also underscores the steep economic costs of forsaking research and development that would bring new green technologies online (Stern 2006).

"Science Fiction" Fictions

Under the title "Science Fiction," Seymour's "Executive Summary" asserts "plenty of evidence—scientific evidence" for her denialist position. Subsequently, her report cites at best scant evidence that could be characterized as under the rubric of science, while ridiculing the "so-called 'science'" (2011: i) cited in the documentary *An Inconvenient Truth* (2006, Dir: Davis Guggenheim) that featured Gore.

Seymour rhetorically summons Dr. Nils-Axel Mörner and endows him with credibility as the former chair of a professional sea level study group (2011: 2), although he currently secures his most receptive audiences in right-wing journals of opinion such as *The Spectator*. Outside Seymour's discourse, Mörner's conclusions about sea level decline around the Maldive Islands have been vigorously confronted by other researchers (Woodworth 2005). As for Mörner's former scientific affiliations, the leadership of the organization that he once headed has strenuously distanced itself from his views on climate change:

> "Dr. Mörner currently has no formal position in INQUA [International Union for Quaternary Research] and I am distressed that he continues to represent himself in his former capacity. Further, INQUA, which is an umbrella organization for hundreds of researchers knowledgeable about past climate, does not subscribe to Mörner's position on climate change. Nearly all of these researchers agree that humans are modifying Earth's climate, a position diametrically opposed to Dr. Mörner's point of view." (John J. Clague, quoted in Payne 2011)

Mörner's credibility has been further damaged for his advocacy of the parapsychological phenomena of "dowsing," as he claims the ability to discover hidden water supplies and metals with a special stick (Lynas & Monbiot 2011). Seymour's other sources include a trio of television weather men, the right wing opinion journal *National Review*, and a book from the self-publisher Lulu.com.

Seymour also appeals for support from the Heartland Institute. In a press release in 2012, Heartland asserted:

> "The people who still believe in man-made global warming are mostly on the radical fringe of society. This is why the most prominent advocates of

> global warming *aren't scientists. They are murderers, tyrants, and madmen.*" (emphasis added, quoted in Hickman 2012: 1)

Heartland thumbs its nose at the fact that, by the numbers, the IPPCC's fourth report was shepherded into form by a global team of 152 lead authors, 26 review editors, and 498 contributing authors; and that it cited more than 6,000 peer-reviewed articles, while having been vetted through more than 30,000 comments and by 625 reviewers (Intergovernmental Panel on Climate Change 2009). Despite the efforts to annul documented reality in favor of visions of "murderers, tyrants, and madmen" authoring climate change papers, Heartland is another gold-plated fount of information for MRC.

As for making a case against Gore's film, Seymour cites a report in *The Times* of London. In doing so, she extends MRC's refusal of original sources and betrays dependence on the news media of which MRC is an ostensible watch dog. Seymour quotes the right-leaning *Times*' strongly misleading claims that "the High Court of London ruled there were 'nine significant errors' in the film and decided that the film's 'apocalyptic vision' was 'politically partisan and not an impartial analysis of the science of climate change'" (2011: 3). The flak lawsuit to which Seymour refers sought to block the Secretary of State for Education and Skills' plan to distribute copies of *An Inconvenient Truth* to all the UK's secondary schools. Although Seymour's account does not suggest this, Justice Michael Burton's decision affirmed the premises of the documentary and enabled the Secretary of State to distribute the copies of the documentary.

More specifically, Burton states that, "I have no doubt that Dr. Stott, the Defendant's expert, is right when says that: 'Al Gore's presentation of the causes and likely effects of climate change was broadly accurate'" (2007: 10). Burton's judgment also states that, "the High Court has made clear [that] the law does not require teaching staff to adopt a position of neutrality between views which accord with the great majority of scientific opinion and those which do not." That is, the Justice posits climate change as a settled scientific explanation that accounts for observed evidence. In contrast with the inaccurate description Seymour culled from mainstream media, Justice Burton explicitly rejects the claim that Gore's presentation was "politically partisan" defined by statute to be, for example, campaigning for a party during class session. In an extended passage, Burton observes that non-partisan politics regularly arise in education (2007: 5-8). In this vein, teaching about World War II necessarily adopts a political stance—it would be stifling if it did not—and may, at the same time, be obviously distinct from partisan (*e.g.*, pro-Labour Party) discourse.

In the interest of accuracy, Burton's decision ordered that nine contentious claims made in Gore's version of climate change science should be identified in the "Guidance Note" that had already been prepared for teachers. In determining where Gore had extended beyond the known evidence, Burton uses the IPCC's fourth assessment report in 2007 (released one year after the documentary) as the benchmark for established scientific consensus (2007: 14). The fact that the IPCC is positioned as the "gold standard" for understanding climate change rubbishes Seymour's insinuation that Burton's decision repudiated climate science; even if the mere fact of an unsuccessful court case did succeed in generating fodder for further flak via sloppy and inaccurate second- and third-hand accounts of the decision.

The Mystery Mogul—or Offers of Flak That Can Be Refused

It is not difficult to harbor ambivalences about George Soros, whose "day job" as a financier has had some convulsive economic impacts (Greider 1997: 238-248). However, the paradox of Soros is that "one of the world's most active and influential anti-communists (...) not to mention one of its most successful capitalists" is regularly harassed by the right (Welch 2003: 1). MRC participates in flak against Soros with two reports during the sampled time period: *George Soros: Media Mogul* in 2011 and *George Soros: Godfather of the Left* in 2012. Salvos of flak begin with the titles of the reports that I will address in chronological order.

Media Mogul presents as a strange title since Soros does not own any media properties, although he could buy or start them up at leisure. The timing of Dan Gainor & Iris Somberg's report of 15 August 2011 coincides, however, with a summer of legal troubles for the empire of an unquestioned mogul (Murdoch) that included resignations, arrests, and the abrupt closure of the UK's *News of the World* despite its brisk sales.

Gainor & Somberg articulate Soros and his Open Society Foundations to a line-up of right-wing obsessions: "pro-abortion, pro-illegal immigration, pro-national health care, pro-drug legalization, pro-big government, anti-Israel and, ultimately, anti-American" (2011: 1). In an ominous register, Gainor & Somberg posit Soros as having "direct ties to more than 30 mainstream news outlets" (2011: 1). Despite assurances that "It's a connection hard to deny," the assumed nexus is notably tenuous on the evidence that Gainor & Somberg marshal (2011: 1). To wit, mainstream media employees also sit on boards that receive funds from Soros, a matter that that Gainor & Somberg problematize as "conflict of interest" while avoiding specifics. By contrast, massive audiences are reached by media channels that are, in fact, owned by aggressively political, interventionist right wingers; examples include Clear Channel Communications (Farrell 2004; Krugman 2003) and Sinclair Broadcast Group (Boehlert 2004, Lieberman 2004).

Beyond inventing rationales for characterizing Soros as a media mogul, Gainor & Somberg reference his support for the San Francisco-based Tides Foundation. Following a series of Glenn Beck's Fox News screeds on Tides as a Soros sock puppet, 45-year-old Bryan Williams set out to terrorize the foundation. The attack was stymied because an intoxicated Williams was pulled over by police for erratic driving en route. He then waged a 12-minute gun battle with police. Following arrest, Williams explained that his intent was "'killing people of importance at the Tides Foundation'" and that Beck's tirades were the inspiration (Hamilton 2010: 1). Writing in *Politico*, Tides founder and chief executive Drummond Pike claims that the foundation receives "far less than 5 percent of our $112 million" in funds from Soros (2010: 1)—and that he had never even met the financier/philanthropist prior to Williams' foiled plot. Even after these ugly events, Gainor & Somberg channel the bizarrely conspiratorial Tides Foundation meme minted by Beck (and repeat it in another MRC report [2012: 7], also echoed by MRC's Seymour [2012b: 2, 3]).

Gainor & Somberg also posit that the "Soros Empire" (2011: 7) of media that he does not own is dedicated to the demolition of Fox

News (owned by a genuine, often hands-on mogul). "When Soros was criticized by Fox," Gainor & Somberg write, "multiple pieces of the Soros empire responded." Gainor & Somberg fume that, "Jonathan Schell, a fellow at The Nation Institute, another part of the Media Consortium, made Fox News out to be anti-Semitic for criticizing Soros" (2011: 7). In Gainor & Somberg's account, the criticisms of Soros to which Schell refers are an anodyne matter, except to the thin-skinned billionaire and his lackeys, risibly playing the anti-Semite card. In fact, on Fox News, Beck accused Soros of having been a Nazi collaborator during World War II. Soros, who is Jewish, was taken in by a Catholic family in his native Hungary; protection that enabled the young Soros to survive to the age of 13 at the end of the war. Among other Jewish voices, the right-leaning Anti-Defamation League's Director Abraham H. Foxman unequivocally denounced Beck's outbursts for being inaccurate, "offensive," "horrific," and "repugnant" (Anti-Defamation League 2010). Schell adds that, "Beck falsely charges that Soros has instigated coups abroad while implying that he plans to carry one out in the US" (2010: 2). Gainor & Somberg's bland characterization of the episode as "criticism" that prompted what they imply to have been a vengeful campaign orchestrated by Soros makes an ugly contribution to the flak genre. It also exhibits the flak tendency of going aggressively on offense while disingenuously assuming the pose of defense.

(Not) Brave, (Not) New, (Not Soros' New) World (Order)

MRC published *George Soros: Godfather of the Left* 10 months after *Media Mogul*. In *Godfather*, Gainor & Somberg all but forget the earlier "Soros as media mogul" hysteria to discourse on Soros-backed foundations. They charge that Soros has "helped foment revolutions" while citing no recognizable examples or evidence (2012: 1). Just as *Media Mogul* seems to present an indirect reply to the Murdoch scandals, *Godfather* appears timed to shout over mounting criticisms of the Koch Brothers' checkbook interventions in US politics (Meyer 2010).

In one line of flak, Gainor & Somberg write, "While Soros has even been nominated for [sic] Nobel Peace Prize, many governments view him as the enemy" (2012: 6). For support, they cite the ring leader of the alleged liberal media gang that MRC is ostensibly tasked with auditing. Specifically, Gainor & Somberg quote from *The New York Times* as follows:

> "In Albania, Kyrgyzstan, Serbia, and Croatia, Mr. Soros's foundations have been accused of shielding spies and breaking currency laws. His employees have been assault [sic] and threatened with imprisonment of financial sanctions for alleged crimes," wrote the *New York Times*. (2012: 6)

Gainor & Somberg clearly align sympathetically with these governments against a figure, Soros, whom they present as a covert dictatorial aspirant. However, the opening two sentences of Judith Miller's article in the *Times* that Gainor & Somberg cite clearly stress harassment toward Soros' foundations by heavy-handed governments. Miller writes,

> For the past decade, George Soros, the Hungarian-born financier and philanthropist, has spent more than a billion dollars promoting a free press and political pluralism abroad—everything the world's authoritarian rulers despise. Now some of these political leaders are fighting back. (1997a: 1)

After thusly priming the reader, Miller marshals further evidence that the Soros supported groups are being flacked in a region that had yet to fully emerge from its authoritarian (in many cases Soviet) past.

In the same article that Gainor & Somberg cite, and following sources in country, Miller characterizes Belarus' regime as assaying to transform the nation into a "'Soviet themed park'." The nation is described as ruled by "'the new face of dictatorship in Europe'" via Alyaksandr Lukashenko's "rule by decree" (1997a: 1). In another of her articles from the *Times* that Gainor & Somberg audaciously and falsely present as fashioning criticisms of Soros' international projects, Miller characterizes Lukashenko as "a flamboyant former collective-farm boss who has spoken admiringly of Stalin and the virtues of dictatorship" (Miller 1997b: 3). Miller's characterizations of Belarus were corroborated contemporaneously with her reporting (Amnesty International 1997) as well as in subsequent years (Amnesty International 2007). Nonetheless, in *Godfather*, Gainor & Somberg slice the reporting down to two sentences and invert the narrative thrust that is evident under any competent reading of Miller's articles. To wit, Soros' foundations are exhibiting courage in fostering the institutions of independent civil society in inhospitable places. In doing so, MRC crudely teases and tortures the *Times*' text to elicit the "answers" that are demanded by a line of flak toward Soros. MRC simultaneously positions itself on the side of neo-Soviet autocrats.

Godfather exhibits further dishonesty. Gainor & Somberg assert that a chapter of Soros' book, *Opening the Soviet System* (n.d.), manifests his desire to impose *Brave New World* dystopia that they call "twisted" (2012: 7). Their report concludes as follows:

> Soros criticized [Aldous] Huxley's work, but it's as if he [Soros] used it as a model for his charitable contributions. Imagine if someone had read George Orwell's *1984* and tried to make it happen. That's what Soros has done. (...) Soros has spent hundreds of millions of dollars funding a *Brave New World* for Americans and even he admits it won't turn out well. (Gainor & Somberg 2012: 8)

Putting aside MRC's embarrassing conflation of Huxley's and Orwell's versions of dystopia (Postman 1986), to believe this claim on its face requires a suspension of one's intelligence. To wit, Soros made his book available on the internet—a curious place for a high profile figure to deposit an alleged blueprint for a drive toward global miserablism. By contrast, a competent reading of Soros' chapter construes it as elaborating his concept of an Open Society; to wit, a social order of citizens empowered to make their way, largely without the encumbrances of tradition that had long dictated a person's place. At the same time, Soros inventories some downsides of the strongly individualistic, contract-driven system, even as they are of a piece with his commitment to capitalism. In particular, Soros cautions that a social order that revolves around individual liberty, largely unconstrained by tradition, can be alienating for its subjects and lack moral ballast. In Gainor & Somberg's rendering of Soros' (at times obtuse, hardly new, but interesting) theorization, the measured qualifications around his concept of Open Society constitute a roadmap to totalitarian dystopia. Unable to criticize Soros' writing on the Open Society for what it manifestly endorses, Gainor and Somberg crudely and dishonestly mischaracterize it for flak purposes.

Although a travesty of scholarship, Gainor & Somberg's MRC report was reposted on a number of other right-wing websites that include www.glennbeck.com. In turn, the comment section of *The Blaze*'s re-posting of *Godfather* generated more than 250 reader comments within days. Comments compare Soros to Satan and Hitler, while many express thinly veiled desires for violence. The comments section illustrates that when it employs a "say anything" method, flak agitates the emotions and prepares credulous readers for commitment to the cause.

Conclusion

The previous chapters of this volume employ Herman & Chomsky's model to demonstrate that the economic and professional structures of journalism palpably, if unwittingly, condition and compromise its pursuit of truth. It is wholly conceivable in the lights of the Propaganda Model that journalists' good will and professionalism will be regularly compromised via the structuring structures in which they labor. However, the journalistic professionalism that at least assures adherence to a procedural form of objectivity is not in play when the alpha and omega of an organization is to generate flak.

While a feature of the propaganda model, flak presents its own particular logics. In this view, MRC as a flak-oriented organization is distinct from incumbent news media in two non-mutually exclusive respects. First, as the model emphasizes, flak constitutes an organized means to unabashedly push incumbent media to encode news discourse in closer alignment with MRC's right-wing orthodoxies. At the same time, MRC-style flak makes demands for ideological purity that can never be realized any more than one can walk to the ever-receding horizon—thereby enabling further indignation and flak-driven demands. Second, MRC's flak discourses may also shape and reinforce the decoding strategies of its dedicated audience. In this view, and regardless of what mainstream journalists actually report, the devotees of MRC flak will decode incumbent media as always already captive to left-leaning ideologies and shadowy entities beholden to "Them."

It also bears mention that MRC distributes its flak wares most readily via internet. Among new media's myriad effects, the enablement of flak is among the most notable in the framework of the Propaganda Model. Elaborating on Evgeny Morozov's skepticism about new media (2011), one may recall the proverb, "A dog barks—and a hundred more bark at the sound." So it is at times on the internet. Something gets posted—and then re-posted in chain reaction, endowing claims with an aura of truth value via sheer repetition, rather than for their rigor. With the rise of new media, flak campaigns—climate change science has a veritable legion of critics! Soros has the media in his pocket!—re-circulate across dispersed media channels, endowing them with the legitimacy conferred by an unorchestrated chorus. Such an information environment is strongly enabling toward flak since even the most contentious claims can be readily found and then endlessly echoed. At the same time, despite evident downsides, new media may also pull at some of the joints that hold the Propaganda Model's together. In light of the stakes, unwinding some of the paradoxes and tensions embedded within new media is the subject that animates the next chapter.

Notes

[1] Penn State exhibited less diligence in other cases. During the same decade, the university's police department and administration was furnished with an eyewitness report of retired assistant football coach Jerry Sandusky sodomizing a ten-year-old boy in one of the campus' shower facilities. In contrast with the immediate investigation of Mann in which due diligence was clearly the rule, Penn State did nothing to perturb Sandusky or to pursue justice. Long after his series of crimes were committed, and due to action initiated outside the university community, Sandusky was finally convicted on 45 counts of child abuse in June 2012.

[2] Although MRC's Graham postures as an ardent warrior advocate, this stance is subject to the situational demands of the double standard. Thus, Graham takes offense that, in a *60 Minutes* interview, perpetual losing candidate for office Willard "Mitt" Romney (born 1945) was asked why he had not served in Vietnam. Graham gripes that it is "a question never put to Obama," born 1961 and

thus all of seven years of at the time of the Tet Offensive (2010: 11). Graham also disapproves when Romney was queried on *60 Minutes* as to why none of his five adult sons compiled a military record, even though they were all around prime fighting age (born between 1970 and 1981) when Bush bravely issued the clarion call for a generation-defining "War on Terror" in 2001. Contra Graham's sudden conversion to pacifistic indignation, these are questions that should be asked of the pro-war elites like Romney who regularly demand ineffable sacrifices from other people and other people's children.

[3] As Mann (2012) observes, many investigators in the 1970s believed that the planet could cool due to industrial pollution partially shielding the earth from the sun (as demonstrably occurs with volcanic ash). However, by the 1990s, evidence-driven consensus moved massively toward the paradigm of aggregate warming on a global scale.

CHAPTER 7
"Eye Rolling" and Rolling Over: Self-Reflexive Criticisms of Journalism in New and Old Media

This chapter orients toward new media developments such as blogs that were unanticipated when the Propaganda Model was formulated. As such, new media may appear to be situated outside the model. However, examination of new media through the model's concepts may furnish a more clear-visioned idea of what is—or is not—new in the ongoing "Communications Revolution." In line with this volume's abiding concern with news media, the current chapter develops a paired case study of how old and new media self-reflexively discourse on the journalistic environment of which they are a part. The case study features Howard Kurtz, a prestige representative of incumbent media from his former position as media writer at *Washington Post*, and Glenn Greenwald's erstwhile blog at the internet daily *Salon*.com.

A high-stakes skirmish is currently being waged between enthusiasts of new media who posit it as an empowering paradigm shift and the doubters who are wary of internet hype. Doubters argue that, while providing surface-level technical fixes to deep-rooted problems, new media even enables undesirable outcomes such as stepped-up surveillance (Morozov 2011). Moreover, doubters maintain that the Communications Revolution intensifies already documented defects of incumbent media such as time pressure on journalists ("churnalism") and the concentration of audiences around a tight circle of market leaders (Curran 2011: 111-120; 2012). Some scholars take the hyperbolic route and claim that new media rewires our brains and makes us "flatter," less like the humanistic glories of a cathedral-like edifice (Carr 2008). Others yearn for more space in which to reflect on political and social life without horizontal pressures to be plugged in (Sterne 2012).

Ranged against new media skeptics who at times broach dystopic visions are the enthusiasts who on occasion raise the specter of utopia. Even prior to the popular ascendency of the internet, many of the arguments for new media (as hipper, more conversational and responsive) were being rehearsed (Katz 1993). In new media, the enthusiasts find cascades of pleasure and proliferating texts to stimulate the mind (Jenkins 2008), information that extends to the horizon (Welch 2011), and more inclusive participation that reconfigures the media terrain (Gillmor 2004). In the enthusiast paradigm, new media generates the platforms for improved (pro-) social organization (Shirky 2008). Some particular new media firms have been lauded at book length, often with aggrandizement toward the presumptive liberationist energies of capitalist entrepreneurialism encoded between the lines. Examples include David Kilpatrick's *The Facebook Effect* (2010) and Randall Stross' *Planet Google* (2008).

All of the participants in this quick survey of discourse on new media are astute enough to, in some degree, qualify their claims. Hence, doubters concede many upsides of new media such as rapid access to vast archives of information. For their part, enthusiasts steer around the charge of vulgar celebrationism by nodding toward "digital divides" and admitting to some stubborn realities that refuse to yield to tweeted imperatives. In preface to the Kurtz-Greenwald paired case study at the core of this chapter, I will mainly accent arguments crafted by the new media doubters. In the case study, however, I will stage a confrontation between highly regarded exemplars of new and old media tendencies that comes down in favor of the former over the latter.

Why chance a chimerical or confused message in this fashion? In examining the evidence, there is good reason to be wary of the chorus of hallelujahs that surround new media. The seductions of this chorus are, of course, cross-examined in popular and scholarly discourses—just not often or rigorously enough, in my judgment. At the same time, one must recognize what is different about new media. While new media has not changed the contours of pre-existing realities, to regard the future as an already scripted repetition of the past is an unimaginative and disempowering posture that courts political paralysis. In this view, small changes can cascade into larger changes; failure to notice these possibilities signifies that *They* have already prevailed, even in one's imagination. Hence, in this chapter, I will elaborate the skeptical view on new media that rejects carnival barker hustle. I then pivot toward the paired case study that engages with hope for blogging that evades the conditioning of the Propaganda Model.

Curb Your Enthusiasm

One of the most bracing critiques of new media has been fashioned by self-described former apostle Evgeny Morozov (2011). He posits that unconditional faith in the liberating impact of the internet is supported on a number of pillars of conventional wisdom that are wobbly on inspection. These pillars include a tendency toward determinism that endows technology with its own essence and motive force. At its crudest, technological determinism fixates on the tool and ignores the environment in which it operates. To take one example, new media maven Dan Gillmor serves up made-to-order techno-determinism when he professes "no doubt that technology will eventually win" and vanquish old media "because it is becoming more ubiquitous" (2004: 238). Determinist assumptions of this sort—"*technology will eventually win*" because of what "it" is "doing"—endow machines with independent will and gloss over technology's immersion within the pre-established culture. To wit, how is the technology produced? Who benefits as a technology circulates and who loses out from it? How is it disseminated, who has access, how is it regulated (Winston 1995)? In critiquing the fixation on a new media instrument's putative characteristics over the exigencies of its place and time, Morozov also perceives a failure to appreciate the resilience of prevailing regimes of authority. These regimes, including repressive ones, assay to reverse-engineer and appropriate new media to serve their needs, in defiance of assumptions about new media's presumed liberationist "essence."

Morozov skeptically appraises the widely maintained nostrum that "doses of information and communications technology are lethal to the most repressive of regimes" (2011: xii). For Morozov, current assumptions about the internet replay tropes of the Cold War as having been clinched by Western infiltration through information campaigns. He argues that the Soviet regime may be construed as having imploded due mainly to internal stresses—a stagnating economy, submerged nationalities—and not as a consequence of information campaigns that are assumed to be a foolproof, non-lethal weapon in Western-style democracy's arsenal. In rebutting such assumptions, Morozov warns that authoritarians may employ new media to play "defense" for the regime via surveillance and sophisticated censorship systems. Regimes may also go on "offense" via propaganda campaigns that, for example, privilege preferred bloggers.[1]

In a discussion that goes some distance down the same path as Morozov' analysis, Rebecca MacKinnon (2012) dissects new media in the People's Republic of China (PRC). She characterizes the PRC as a paradigm for a regime that has successfully surfed the new media wave. The sheer size of the PRC helps on this score since it has been able to develop lively indigenous websites and services (RenRen, Kaixinwang) that keep its population captivated within the virtual boundaries of the nation.

MacKinnon also discusses Chinese citizens who would have been victims of abuse (*e.g.*, unjust imprisonment) but were spared by new media interventions and the publicity channeled through them; moments that evoke the letter-writing publicity campaigns long practiced by Amnesty International. In a related vein, MacKinnon claims that the internet has enabled the PRC's government to forge more "personal" relations with citizens by encouraging participation and practicing responsiveness in electronic fora. Hence, the regime facilitates online discussion about policies that are negotiable (for example, the future of the "one child" policy). Some better governance has resulted. Beyond these internet-enabled advances that dovetail with the regime's interests, lines are clearly delimited with respect to policies that are not negotiable in the PRC (for example, one party rule). Activism outside these lines may generate heavy-handed responses.

The PRC's strategy on managing the "untamed" internet demonstrates that the regime seeks to shepherd new media toward the form that suits its needs. A leaked presentation to top PRC leaders traces a strategic grid as follows (2012: 39):

> "We are following the overall thinking of combining Internet content management with industry management and security supervision; combining prior review and approval with supervision afterwards; combining technological blocking with public opinion guidance; combining hierarchical management with local management; combining government management with industry self-regulation; and combining online monitoring with offline management." (quoted in MacKinnon 2012: 39)

What are some of the more specific ways that the PRC's regime channels the internet in its preferred directions? Rigorous censorship blocks any unwanted content from the rest of the world through the techno "Great Wall." Other unwanted content is not blocked but harassed—slowed down and riddled with connection errors—while content on favored web pages is furnished via free broadband to the elite-in-training at the universities. "Astroturfing" is orchestrated in which faux grassroots (hired) voices speak up for the government line. However, fakeness and simulation are not enveloped in a lonely struggle: "Beyond the Astroturfers, there are also many young patriotic Chinese who create online communities on their own initiative" to, for example, confront perceived anti-Chinese bias from abroad (2012: 45). These are the true believers whom the regime can position to be heard. In contrast with the standard assertions about informational liberation, MacKinnon concludes that, "unless Chinese entrepreneurs and CEOs decide that it is in their interest to build a different kind of future, [it is] more than likely [that] the Internet's pervasive use in China will actually help prolong the Communist Party's rule rather than hasten its demise" (2012: 50). In all of this, MacKinnon pointedly observes, Western technology firms have often been complicit and enabling.

Blogs: Lighthouse or Fog?

As for the already "liberated" West, how does new media play out? At book length, Eric Boehlert (2009) lauds bloggers whom he appraises as having transformed the news media ecology. Out-of-nowhere blogger success stories that he covers include John Amato (*Crooks and Liars*), Markos Moulitas Zúniga (*Daily Kos*), and Greenwald (*Unclaimed Territory* and later *Salon*.com). Boehlert claims that bloggers have satisfied a need for new voices uncompromised by mainstream media's long retreat into complacent insularity. Brigades of motivated, volunteer bloggers scored numerous triumphs of investigation and commentary in their inaugural years on the beat—and even as incumbent media largely dismissed them as "polarizing, amateurish extremists" (2009: xiii). Triumphs include having vetted "Sarah Palin better than the GOP (Republican Party) had" in 2008 (2009: xi). Boehlert credits left-leaning bloggers, perhaps hyperbolically, with grafting "a spine to the Democratic party"—and with having "literally kept the lights on during a very dark period" of George W. Bush's tenure (2009: xvi). These talented and even iconoclastic writers would formerly have depended upon syndication in newspapers and magazines that may, in turn, have no interest in harboring iconoclasm. The advent of blogs should be welcomed by all who desire a more robust public sphere.

However, what does Boehlert's in-depth attention to blog success stories overlook? In important respects, bloggers cannot wholly escape the orbit of prevailing culture and its political economy. In this vein, C. Edwin Baker acknowledges that "reduced distribution costs lower a significant barrier to entry" and "enable a dramatic increase in opportunities for noncommercial and voluntary noncommodified content creators" (2007: 101). In the case of the internet, the most costly barriers to producing a media entry, such as newsprint, have been mooted. Eschewing techno-mania, Baker nonetheless emphasizes that a high-tech gloss on an established activity is not necessarily better than the old-fashioned version. Moreover, as concerns quality, there are famously few guarantees of it on the internet. Baker compares internet content to Wal-Mart, where mountains of cheap wares that have been produced elsewhere accumulate under one roof.

Going by the numbers, Baker also finds that "concentration of audience attention is extreme" in the blogosphere (2007: 107). A star system may indeed be endemic to mediated communication in a class striated mass society and will not be readily arrested by new technical platforms. Consider that, during the Bush era, the US' most read blog was *DailyKos* with an impressive 642,520 daily visits (Baker 2007: 108). This is a noteworthy accomplishment for a medium that did not even exist ten years earlier. It also means that a new media entry emerged out of the grass roots with reach comparable to that of incumbent media. However, drop-off in blog traffic is dramatic after the "super star" bloggers. The eleventh-ranked blog attracted 102,896 daily visits, and the one-hundredth-ranked blog logged 8,410 (Baker 2007: 108). Moreover, the necessary new media accoutrement of search engines tends strongly to reinforce the position of the highest rated sites. To not make the first page of search results is often to be effectively invisible.

Baker's contrast between blogs and US newspapers is also instructive. *USA Today* is the industry leader with daily circulation of over 2 million. Much further down the scale, the seventy-fifth-ranked paper (*Des Moines Register*) boasts circulation of 154,885. Moreover, as the same copy of a newspaper is often read by multiple persons, *USA Today*'s and the *Register*'s hard-copy readerships are assumed to respectively exceed 7 million and 300,000. In this comparison, the blogosphere is characterized by both fewer readers than newspapers—*and* a more consolidated star system of concentrated readership. Incumbent media such as *USA Today* and the *Des Moines Register* also blow new media away on the new media's own terrain. The numbers indicate almost 10 million daily visitors to *USA Today*'s website and over a quarter million for the *Des Moines Register* from its base in a small city (Baker 2007: 108). At the same time, "over 99 percent [of blogs] will be lucky to receive one visit" on a given day (Baker 2007: 109).

Following their heady entrance into the mediascape in the past decade, many enduringly successful blogs have come to resemble daily magazines with a dedicated staff (for example, *Daily Kos*) or have secured a niche on a larger platform (Greenwald and *Salon*.com). In each case, new media comes to resemble old media a little more. Baker adds that, "major Internet sources mostly have the same owners that, according to critics of media ownership concentration, were too concentrated before—and are now too concentrated irrespective of—the Internet" (2007: 113); a situation further hardened into place by cross-subsidization of online property by revenue and content from the large firms' incumbent media holdings.

Surveying developments in culture and political economy, Hesmondalgh (2007) enunciates a position that presents useful nuance. He writes that, while "The Internet is full of material," its "many minor forms of subversion, insubordination, and skepticism don't cancel out the enormous concentrations of power in the cultural industries"; nevertheless, the internet "might be thought of as representing a *disturbance*" (original emphasis; 2007: 262). While the current arrangement is neither immutable nor immortal, "the radical potential of the Internet has been largely, but by no means entirely, contained by its partial incorporation into a large, profit-oriented set of cultural industries" (Hesmondalgh 2007: 261).

Howie versus Glenzilla: Howard Kurtz' Corpus of Articles

This chapter revolves around a paired case study in new and old media in their self-reflexive discourses on journalism. Articles were harvested from a Lexis Nexis search of "Howard Kurtz" as author between the dates 1 January and 1 July 2010. The search yielded 52 articles from the *Washington Post* that totaled 55,400 words (an average of 1,065 per column). The analysis of Greenwald was developed from 40 posts that he made on the *Salon*.com internet daily during the same six-month period in 2010, mainly on the same days on which Kurtz published an article. The sample of Greenwald's posts totaled to about 59,000 words or slightly less than 1,500 per post.[2]

Despite all of the pre-standing factors that continue to constrain new media, considered earlier, Kurtz' and Greenwald's discourses on journalistic performance play out as a new media enthusiast would predict. Greenwald exhibits free-wheeling analysis that is less wedded to State and private sector sources and their favored narratives. By contrast, as a mainstream journalist tasked with writing on journalism, Kurtz enthrones conventional wisdoms and the authority that animates them. In doing so, Kurtz confronts what are taken to be "extremist" deviations among some journalists while extolling standard concepts of objectivity. However, despite the pronounced differences between the two writers, new and old media cannot be assumed to form a strict dichotomy. As if it to underscore the point, in October 2010, Kurtz left the *Post* after 29 years to take up a position at the internet *Daily Beast* for what was reported to be a huge pay envelope (Blodget 2011). While this chapter was being drafted in August 2012, Greenwald announced his departure from *Salon*.com to blog from the platform of the UK *Guardian* (Greenwald 2012).

Table 7.1: Sampled Articles from Kurtz and Greenwald

Howard Kurtz: Dates of Published Writing
January 7, 9, 11, 12, 13, 15, 18, 22, 25, 30; February 1, 8, 15, 22; March 1, 8, 10, 15, 22, 25, 26, 29, 31; April 3, 5, 7, 12, 13, 16, 19, 26 (two articles), 30; May 3, 5, 6, 7, 10, 24 (two articles), 31; June 7, 8, 10, 14, 21, 24, 25, 26, 28, 30; July 1.
Glenn Greenwald: Dates of Published Writing
January 7, 9, 11, 12 (two posts), 15, 18, 22 (two posts), 26, 31; February 1, 8, 15, 22; March 10, 15, 22, 25, 26, 31; April 6, 7, 13, 17, 19, 26; May 10, 24, 31; June 8, 10, 14, 21, 25, 26, 28, 30; July 1.

Ruling the Pond

The selection of Kurtz for this case study is given by his status. Graff writes, "In an age when just about anyone can be a media critic, one fish is still the biggest in the pond" and "the nation's preeminent media reporter" (2005: 1). Kurtz' early work in journalism was at an oppositional student paper at State University of New York at Buffalo and on investigative reporter Jack Anderson's staff during the early 1970s. Kurtz was thus inducted into the profession at a moment when the zeitgeist of Watergate and counter-culture wafted through it. However, in time, Kurtz gave up the "unshaven counter-establishment" posture for a position at the high-prestige *Washington Post* in 1981 (Graff

2005: 3).

In 1990, Kurtz was appointed as the *Post*'s media writer. From this platform, he "is largely credited with establishing the media as a regular day-in and day-out beat and inspiring similar reporters around the country", while being read intently within media industries (Graff 2005: 4). He has also penned five books on media while keeping up with other journalistic duties (Cable News Network 2010). While serving as the media writer for the *Post,* Kurtz performed similar work for CNN on its program *Reliable Sources.* Thus, depending on the day of the week, Kurtz was in position to critique—or refrain from critiquing, even if inclined to do so—one of two major news outlets from the platform of the other.

The Kurtzian Discourse in Overview

In his reports during the first half of 2010, Kurtz writes on a variety of issues that include new wrinkles in journalism. For instance, Kurtz examines ProPublica, an innovative organization designed to partner with established news outlets in conducting investigations that would otherwise be too involved for even prestige incumbent media (2010m). In covering the industry, Kurtz also engages with navel-gazing as several articles are devoted to affairs of *The Washington Post* that employs him (2010r). In reporting on industry trends, Kurtz provides an interesting account of one veteran journalist's work regime after ABC News shed about a quarter of its work force. The reporter has become a "one man band" who investigates, scripts, shoots and edits the reports that he files. Kurtz' objective approach to the story locates up- and downsides to the arrangement. He offers the measured conclusion that, "A one man band is cheaper, quicker, and more nimble—but cannot produce the deeper sounds of a small journalistic orchestra" (2010g: C1).

Kurtz betrays ambivalence about new media, although, as noted, he joined the staff of an internet publication in 2010. These undercurrents gain expression in Kurtz' selection of stories that include a blogger with a history of plagiarism who outed a rumored-to-be-gay public figure. By contrast, he notes, "most major organizations have policies against" outing (2010l: C1). Consistent with suspicions about new media, Kurtz reports on a specious story that famous-for-being-famous Lindsay Lohan had contracted the HIV virus (2010q). Kurtz observes that the fake story was said (also incorrectly) to have originated with her father's Twitter account that incited further internet discourse on the ersatz event. In a more positive register toward new media, Kurtz devotes objective but tacitly affirmative coverage to the newly formed "online news service GlobalPost," taking note of its fleet of reporters in 50 countries (2010j: C1).

Kurtz is attentive to the business dimension of new media's impact: "Online advertising revenue still represents pennies on the dollar compared with print advertising" (2010p: A16). Solving the internet revenue conundrum is presented as a matter of moment, as the article concludes with a 27-year *Newsweek* veteran's anxious observation that "'with the right kind of Web strategy (...) we can start prospering again'" (2010p: A16). Kurtz also damps down the assumption that new media has stirred up news anarchy. One report cites the Project for Excellence in Journalism's calculations that "Eighty percent of the traffic for news and information is vacuumed up by the top 7 percent" of internet sites. These sites are, in turn, operated by the familiar incumbent media and aggregators such as Yahoo (2010h: C1).

Kurtz also discusses new media's intersection with the Obama administration. In his account, "network executives were complaining that the White House was costing them tens of millions of dollars" with prime time press conferences that capture less advertising revenue than regular programming—and so Obama obligingly stopped doing them (2010e: C1). The observation presents stark acknowledgment of the tension between commerce and news. Although Kurtz raises more than explains the issue, he adds that the Obama White House's penchant for "scripted and very on-message" communication tactics may actually be better suited to the new media milieu (2010e: C1).

Reporter on the Beat

Mainstream media is often critiqued for its devotion to assessing performances and postures over substance. In this view, complacent punditry flourishes in the stead of analysis. In his role as media writer on a prestige platform, Kurtz often gives unwitting credence to the critique. For instance, Kurtz devotes almost 1,300 words to the genre of celebrities making public apologies (2010f). The report dwells on Tiger Woods and asks whether the philandering golfer should have taken questions from reporters to better appease them during a public apology for his private behavior. In the same article, Kurtz ponders whether a couch is a useful prop on which to stage a *mea culpa*. Manifestly more important questions that impact news media generate far briefer treatment. Specifically, Kurtz reports that the Obama administration mobilized on the legal front against journalists who report leaks (2010o). The administration was in the process of subpoenaing James Risen who earned a Pulitzer Prize for his examinations of Bush-era domestic surveillance. Despite the salience to the core journalistic mission of investigation, Kurtz exhibits far less curiosity (and devotes half as many words) to this story than toward grading celebrity apologies.

As concerns the substance of journalism's mission, one of Kurtz' articles addresses an embarrassing moment for his employer (and, more broadly, for incumbent media). Kurtz acknowledges that, six months earlier, the *Post* released marketing fliers that promised

> an intimate and exclusive setting *Washington Post* Salon. (...) Bring your organization CEO or executive director literally to the table with key Obama administration and Congressional leaders. (...) Spirited? Yes. Confrontational? No. The relaxed setting in the home of [Post publisher] Katharine Weymouth assures it. (2010b: C4)

The price tag for the newspaper's pimp-like services was up to $(US)25,000. For this institutional *mea culpa* story, Kurtz' characteristically objective idiom enables him to dead-pan through the embarrassing revelations. The relative reticence of 631 words in the obscurity of page C4 helps on this score.

Nonetheless, the *Post*'s eventually scrapped plan for influence peddling may not be as revealing of mainstream media practices as some of Kurtz' other efforts. He often uses his platform to unselfconsciously valorize journalism's pack behavior and to recycle conventional wisdom. In this vein, Kurtz' discussion of the Obama administration's health care bill exhibits striking complacency about news media. While claiming that the press took a "bum rap" for not educating and often confusing audiences, Kurtz proceeds through a litany of alibis (2010i: C1). He insists that journalists were "overwhelmed," distracted by insider chatter and other ephemera. As a result, "months and column inches [were] wasted on Max Baucus and the Gang of Six," the general vicissitudes of Obama's fortunes, and the siren call of "the latest Tiger mistress to go public" (2010i: C1). Kurtz' post-mortem on coverage of the health care bill traffics in glib punditry and reads like a fraternity house's self-appointed "historian" musing on the narrative arc of the previous night's blow-out party. One might not realize from Kurtz' account that the Affordable Care Act was one of the most significant bills in decades and proved sufficiently contentious that only a Supreme Court decision in 2012 cleared the path to its implementation.

The corpus of Kurtz' articles include an appreciable amount of fluff that is alien to Greenwald's posts. Kurtz' puff piece on journalist Chuck Todd brings a series of dubious prevailing media values into focus, if inadvertently. Todd is monumentalized as a savvy contemporary media player, with "street cred'" bestowed by "his goatee, infectious grin and steady stream of pop-culture references" (2010s: C1). Todd's penchant for exposure is satisfied through platforms at NBC and MSNBC as well as on Twitter and in the blogosphere. The substance of Todd's work, as channeled through Kurtz, presents an unimpressive *bricolage*. Todd rates performances (the US Ambassador to Afghanistan acts as if in "'a hostage tape'") and serves up middle-brow cliché ("'Americans want candidates, Democrats or Republicans, as angry with Washington as they are'")(2010s: C1). A specific story that Kurtz pulls from Todd's portfolio is "'How to train a hamster'" as the *Post* media writer's puffery is synergized by his subject's soft focus nose for news.

When not puffing other media workers, Kurtz devotes similar attention to the State's agents. In an article on Joe Biden that mainly grades his mediated performances, Kurtz observes that a news magazine's cover story on the vice president was composed with "ample cooperation" from him (2010v: C1). The media writer recounts Biden's "assiduous courtship of the press, hosting a

Saturday afternoon lunch party at his official residence for White House correspondents and playing water tag with their children" (2010v: C1). In Kurtz' reportage, media-State confluence via information subsidy and access are at once revealed—*and* treated as cheerfully as the benign characteristics of the weather.

Policing Boundaries

Kurtz offers bursts of interesting reporting circumscribed within the procedures of objective journalism. The moments when the media writer most clearly abides by the ideological demands of the sphere of consensus arise, however, in articles that address what he takes to be reportage from beyond its perimeter. In these instances, Kurtz writes as if deputized to police boundaries and to enthrone mainstream methods and assumptions.

Kurtz' treatment of the retirement of longtime reporter Helen Thomas in June 2010 illustrates mainstream distaste toward colleagues believed to have hurtled outside the sphere of consensus. It bears mention that Thomas' sudden retirement at age 89 was occasioned by comments in which she instructed Israelis to "go home" to Poland, Germany and the US. Telling any legitimately settled person to "go home" may be unavoidably construed as offensive.

In covering Thomas' inglorious exit from the profession, Kurtz piles on to claim that "colleagues rolled their eyes at her (Thomas') obvious biases" when she was an active journalist. He quotes one reporter who adds that Thomas "'asked questions that no hard news reporter would ask, that carried an agenda (...) sometimes her questions were embarrassing to other reporters'" (2010u: C1). The examples of such questions that Kurtz cites in this vein are indeed embarrassing, but not in the sense that he presumably means. To wit, one question probed George W. Bush's press secretary Tony Snow on Israel's 2006 bombardment of Lebanon's civil society. With implicit after-the-fact support from Kurtz, Snow dismissed the question as reflecting "'the Hezbollah view'"; an accurate characterization, once one assumes that organizations such as Amnesty International (2006), Human Rights Watch (2006), and the Roman Catholic Vatican (Global Research 2006) are similarly captive to "the Hezbollah view" when they criticize Israeli conduct in the region. As the aforementioned sources also condemned Hezbollah's actions, it is obvious that criticism of Israel's military ventures do not collapse into "the Hezbollah view," as Us/Them chauvinism manipulatively implies.

While Kurtz' 8 June article on Thomas is not wholly unsympathetic, it demonstrates how a dubious view of a journalist's subject matter can be conveyed through the mores of objective reporting and marshaling of quotations. As if to dispel any earlier ambiguities, Kurtz returns to Thomas six days later. The report is garnished with further references to "eye rolling" from her press colleagues and stepped-up ideological policing. Following references to her earlier impact as a female reporter when they were notably scarce, Kurtz takes umbrage with this question that Thomas posed to Bush in 2007:

> "Mr. President, you started this war. It's a war of your choosing. You can end it, alone. Today. (...) Two million Iraqis have fled the country as refugees. Two million are displaced. Thousands and thousands more are dead. (...) We brought al-Qaeda into Iraq." (quoted in Kurtz 2010w: C1)

Kurtz intones that, "One might agree or disagree with *those sentiments*, but she was performing as an activist, not a journalist" (emphasis added; 2010w: C1). For this, Kurtz adds gravely, Thomas failed to "meet certain minimal standards." Kurtz' ideological police action notwithstanding, dispatches from material reality confirm that Thomas' question was not one about which people "might agree or disagree," but predicated on well-documented facts that he patronizingly recasts as "sentiments."[3] Moreover, if for the risk of being dismissed as activists, reporters should only lob friendly questions or ones that can be ritualistically slapped down by authorities, then the profession serves no discernibly useful purpose. Kurtz' call for a pacified press corps also manifestly rejects even the mythology of its antagonistic functions.

Following the ideological police action against Thomas, Kurtz' column glides to the next news item under the sub-topic "Asked and Answered." Without irony, the item suggests the kind of journalism of which the *Post's* media writer apparently approves and that satisfies the rarified standards. The item reads as follows:

> Sarah Palin was right when she told Greta Van Susteren that "you're not afraid to ask the questions." The Fox News host indeed asked a question that was buzzing online—started by photos on Gawker. "Breast implants: Did you have them or not?" Van Susteren said. (Kurtz 2010w: C1)

("Eye rolling"? "Embarrassment"? "Minimal standards"?)

Kurtz' ideological dragnet confronts other extremists alienated from the political center. In doing so, Kurtz behaves in congruence with Gans' ethnographically informed characterization of journalists as devoted guardians of mainstream values (Gans 2004). When covering the Unification Church-owned, aggressively right-wing *Washington Times*, Kurtz maintains the convention of objectivity while he hints that the newspaper is the equivalent of a banana republic (2010a; 2010n). Events that Kurtz reports, such as abrupt turnover in the *Washington Time*'s editorship, give credence to the insinuations.

As concerns extremism, Kurtz also discusses Glenn Beck's troubled tenure at Fox News. He writes that, for some observers, Fox was tarnishing an already dodgy reputation by its association with Beck. The network was also losing scores of advertising accounts, as well as staffers who had clashed with the volatile performer. With clear if implicit disapproval, Kurtz notes that, "Beck sparked criticism from some Christian leaders last week when he urged parishioners to leave churches that promote 'social justice' or 'economic justice' saying those are 'code words' for communism and nazism" (2010h: C1). Left-leaning audiences may find Kurtz' ideological policing more palatable when it sanctions Beck and Fox News. It is noteworthy, however, that Kurtz' devotedly centrist prose about Beck does not approximate the condescension directed toward Thomas—although, as noted, Thomas aligned with facts on Iraq whereas Beck is as inflammatory as he is vacuous.

Kurtz' patrols on behalf of mainstream journalism implicate *The National Enquirer* when its executive editor announced plans to apply for Pulitzer Prize honors. The series of *Enquirer* articles to be nominated concerned former senator and vice-presidential candidate John Edwards' sordid personal life. "Don't laugh", Kurtz counsels in reference to what he construes as the tabloid's hubristic bid (2010d: C1). Haughtiness aside, scholars have long noted continuities between mainstream journalism and tabloids. The mainstream valorizes sources such as "senior administration figures" on background, while tabloids may be more in their element as they approach tarot card readers. Both categories of sources are regularly cited with complete credulousness in, respectively, mainstream and tabloid media (Bird 1990). However, Kurtz assays to differentiate the mainstream from the tabloids as the former "do not pay for information" and is further affronted by the *Enquirer* editor's "lectures on morality" (2010d: C1). In this view, mainstream media has no corrupting entanglements with the subjects that it covers, notwithstanding the impact of sourcing patterns and information subsidies. Despite Kurtz' superciliousness on the issue, elsewhere he reports without irony on tabloid discoveries (2010t).

Kurtz also endows Mark Halperin & John Heilemann's *Game Change* volume with enthusiastic coverage even as it approaches the tabloid-style discourse of which he ostensibly disapproves. Indeed, the same topic that Kurtz sniffs at with respect to the *Enquirer*—Edwards' private life—animates his coverage of *Game Change*. Kurtz recounts the deeply tacky story of Elizabeth Edwards "in an airport parking lot, tearing off her blouse and wailing to her husband, 'Look at me'" (2010c: C1). Following the norms of an insiderly oral culture, Kurtz also participates in guessing games about which of Halperin & Heilemann's anonymous sources furnished which quotes. The speculation is garnished with stout defenses of anonymous sources engaging in what may merely be ax-grinding and score-settling, in contrast with anonymity for the purpose of genuinely consequential whistle blowing.

As noted, Kurtz has been exalted as the premier media writer within the mainstream who reports on the reporting of legitimate controversies. It is perhaps unsurprising that, along with some decent reporting, he is a militant defender of mainstream mores and methods, even where they prove defective to the pursuit of truth.

Howie versus Glenzilla: Greenwald's Corpus of Articles

According to his profile on the *Salon*.com webpage, as of 2012, Greenwald has authored three books that achieved *New York Times* bestseller status. The *Salon* profile notes that he was awarded the inaugural I. F. Stone Award for independent journalism and, in 2010, was recognized with the Online Journalism Association Award for his examination of the US government case against WikiLeaks leaker Private Bradley Manning. While Greenwald is not explicitly identified as a media writer like Kurtz, news media critique is a regular theme of his commentary. It is apparent that Greenwald assumes news media to be a central feature of the superstructure of power, hence there is sufficient common currency to enable comparison of his discourses with those of Kurtz.

As concerns back story, Greenwald was bored with practicing law at the same time that he was becoming interested in blogs that he followed early in the Bush era. In October of 2005, he made his blogging debut with *Unclaimed Territory* (Boehlert 2009: 186). Greenwald's lucidity and insight quickly drew attention and his readership spiked—from 30 to 30,000—in all of five days with a boost from timely links from more established blogs (Boehlert 2009: 186). In the coming months and years, Greenwald's dogged attention (hundreds of posts) is credited with having impacted the discourse on the Bush administration's defiance of Foreign Intelligence Surveillance Act (FISA) laws. "Like a good lawyer presenting a complicated case to a jury," Boehlert comments, Greenwald "constructed his narratives from the ground up, adding in a dose of appropriate indignation as he proceeded, all wrapped in meticulously researched writing" (2009: 187). His vigorous "take downs" also earned him the moniker of "Glenzilla." In 2007, Greenwald's blog moved onto the *Salon*.com platform, itself a new media player of note as the longest-running internet daily that has been online since 1995 (Goss 2003).

Along with being insightful, Greenwald's discourse on contemporary news media occasionally drifts into uncomplicated functionalism. In broad strokes, he writes that it is "not only the effect but the intent of the central method of American journalism: to disseminate outright falsehoods to the American public" (2010d: 1). While falsehood is demonstrably an outcome of US reportorial practices at times, Greenwald takes the additional step of characterizing it as the very purpose of journalism. He thereby places a simplifying gloss on how a complicated and contradictory society's institutions and interest groups behave. Chomsky & Herman's model is far more diagnostic in identifying the different moving parts of the media machine that are independent of the largely well-meaning intentions of media workers. Herman & Chomsky's account suggests, moreover, that the media machine's component parts may move out of sync and even jam on occasion. Greenwald also clearly enjoys indulging in personalization and taking verbal smacks at figures of whom he disapproves. These practices are not to all tastes.

Greenwald nevertheless drills down into particular journalism-salient stories in a manner that Kurtz, an ostensible media specialist, does not. Moreover, Greenwald frequently assumes positions in contradistinction to the conventional wisdoms that Kurtz valorizes and re-circulates. For instance, Greenwald cites a newspaper account of the US Congress' work on a bill to ban US firms from "'working with governments that digitally spy on their citizens,'" a jab at the People's Republic of China (2010e: 1). While there is no need to defend the Chinese regime's plainly illiberal practices, Greenwald eschews Us/Them dichotomizing to point out that Congress' posture toward China is tougher than its more emollient stance on digital surveillance within the US.

Greenwald also performs close (and legally-trained) readings on mediated arguments, such as those made in favor of the Obama administration's nomination of Elena Kagan for the Supreme Court (2010m). In this case, Greenwald focused critically on pro-Kagan testimonies from three new media outlets (*Slate, Huffington Post*, and *SCOTUSblog*) that augment his regular attention to incumbent media. Greenwald also incorporates academic research into his blog posts on occasion. For instance, he details a Harvard University study that found that prestige US media outlets reversed their long-standing practice of almost monolithically describing water boarding as torture after the Bush administration openly embraced the practice (2010s).

The very things that bother Kurtz are what Greenwald lauds—and vice-versa. In January 2010, Greenwald blogged that, "Helen Thomas shows—yet again—that she's one of the very few White House reporters willing to deviate from approved orthodoxy scripts" (2010b: 1). In differentiating Kurtz' and Greenwald's discussions of news media, the case of Halperin & Heilemann's *Game Change* also exhibits contrast with neon clarity. While Kurtz is almost giddy about the book, Greenwald condemns it for channeling the "tone, content, and 'reporting' methods" of the *Enquirer*. Furthermore, the book's pre-released teasers on Edwards pile on a "completely destroyed non-entity with no political future" via "sleazy voyeurism" that makes no contribution of value to public discourse (2010c: 2). In a follow-up post, Greenwald characterizes the book as wholly constructed from "unattributed, shielded sources" enveloped in anonymity and even in paraphrase (2010d: 1). Thus, while both Kurtz and Greenwald train fire on tabloids, Greenwald extends his criticisms toward what he presents as being regularly manifested continuities between yellow and mainstream journalism. By contrast, Kurtz wagers heavily on the rhetorical maintenance of a strict dichotomy between the vulgar gusts of the tabloid and the rarified airs of the mainstream.

Principles First

Greenwald is not beholden to the cramped procedures of the objectivity doctrine that has, nevertheless and in practice, enabled journalists to fly their opinions under the radar. Instead, Greenwald often and openly presents his guiding principles. Commitments include vigorous advocacy for the First Amendment with some touches of iconoclasm (2010f; 2010i). Another fundamental of Greenwald's worldview is the clear separation of distinct nodes of authority in a well functioning society. Distinction from Kurtz is apparent since the *Post* critic seems agnostic to the circulation of power, even as he is supportive of the pillars of prevailing authority, such as journalism's close proximity to the State (as in his discussion of Biden). By contrast, Greenwald diagnoses "a political culture drowning in hidden conflicts of interests, exploitations of political office for profit, and a rapidly eroding wall separating public and private spheres" (2010j: 1). Michael McConnell, the Bush-era Director of National Intelligence and current Vice Chairman at Booz Allen (a sprawling intelligence firm), presents Greenwald's lead example of the power elite. In this arrangement, only the rapidly spinning revolving door stands between State and private authority. Identifying the stakes, Greenwald posits that "corporatism (control of government by corporations) was the hallmark of many of the worst tyrannies of the last century" (2010j: 2).

Greenwald's concern with the dispersion of power is evident in his longest post during the sampled period of approximately 4,000 words. Here, Greenwald makes the "positive" case for what he believes via an approving review of the record of Seventh Circuit Court of Appeals Judge Diane Wood (2010n). He characterizes Wood as making decisions that stanch the concentration of executive branch power that has dramatically intensified in recent US history. Greenwald also depends on original (legal and scholarly) writings in fashioning his portrait of Wood. Sourcing patterns that incorporate scholarly texts present dramatic contrast with Kurtz' reliance on the oral culture of interviews and backgrounders that enthrones sources drawn from the rolodex (Kurtz 2010k).

With respect to US foreign policy in the Middle East, Greenwald argues that the US pursues a self-defeating path as calibrated against stated goals. Following his condemnations of al Qaeda and underscoring that "deliberate targeting of innocent civilians" carries no legitimacy or justification, Greenwald elaborates:

> It's truly astounding to watch us—for a full decade—send fighter jets and drones and bombs and invading forces and teams of torturers and kidnappers to that part of the world [Middle East], or, as we were doing long before 9/11, to overthrow their governments, prop up their dictators, occupy what they perceive as holy land with our foreign troops, and arm Israel to the teeth, and then act surprised and confused when some of them want to attack us. (...) As a result, we have come to believe that any forms of violence we perpetrate on them over there is justifiable and natural, but the Laws of Humanity are instantly breached in the most egregious ways whenever they bring violence back to the U.S. (...). (2010a: 2)

This and many other passages adopt an idiom and a technicolor vivid message not found in the establishmentarian *Post*.

While one may disagree with his polemical moves and conclusions that he offers, it is evident that Greenwald evaluates claims on their merits—and not on their alignment with conventional wisdoms that incubate within the spheres of consensus or legitimate controversy. By contrast, within the mainstream where Kurtz is domiciled, claims are often measured by their fit to the prevailing wisdoms and not to external reality as rigorously apprehended.

Media Critique

On the Kurtzian view, an occasional reporter is an eye-roll inducing activist (Thomas) and one finds a journalistic banana republic here and there (*Washington Times*). Beyond readily identified deviants, and some uncomfortable commercial pressures on news organizations, it is nevertheless "all good in the 'hood" in Kurtz' view. The journalistic system behaves largely in congruence with civics lesson platitudes for Kurtz while it ministers to the legitimate controversies. By contrast, probing media critiques are regularly stitched into Greenwald's posts, given his implicit assumption that all issues reduce in some form to media and messaging in the current ecology.

As concerns domestic policy, Greenwald's background in law frequently comes into play—and often for the purpose of puncturing the conventional wisdom promulgated by media workers and other elites. In one instance, Greenwald critiques a video issued by Senator Susan Collins (Republican-Maine), usually cast as one of the "moderate" members of her party's caucus. Noting that her video message vacillated between "the angriest, most unhinged version of Dick Cheney" and "standard right wing boilerplate," Greenwald telescopes in on one claim that Collins made while assailing the Obama administration (also assailed by Greenwald, but on different grounds). In particular, Collins recycles "one of the most pervasive myths in our political discourse: namely, that the US constitution protects only Americans and not any dreaded foreigners" (2010h: 1).

Greenwald points out that in the 2008 Supreme Court case *Boudmediene versus Bush,* all nine justices agreed that people are subject to a battery of protections on US soil regardless of whether they are citizens. That is not a surprise. Non-citizens (for example, visitors on student visas) who are arrested must by law be charged with a recognizable crime and subject to trial, among other constitutional protections. What has been at issue, Greenwald explains, is how such protections from State power apply beyond the US' borders. Indeed, the Bush administration's rationale for the off-shore Guantánamo facility was to evade constitutional obligations by playing geo-legal shell games. Greenwald also points out that expansive protections for *people* (and not just the sub-category of them who are *citizens*) has been settled law since the nineteenth century. By contrast with Greenwald's forensic thrusts, Kurtz generally invigorates conventional wisdoms while fashioning objective accounts and defending journalistic monuments to the establishment. In following the objectivity doctrine, what Kurtz does *not* do is unpack whether the statements and political positioning that he often evaluates in performative terms are informed by truth.

During the sampled period, one of Greenwald's most direct confrontations with mainstream news media is called "How Americans are Propagandized about Afghanistan"—a title unlikely to animate an article in incumbent news media. Greenwald recounts a February 2010 incident in which four people were gruesomely killed in Paktia province, Afghanistan. As the event was originally reported by prestige outlets CNN and *New York Times,* the culprits were Taliban members who executed honor killings of the bound and gagged victims. In turn, CNN's and the *Times*' reportage originated from NATO spokespeople. The NATO narrative fit like a carefully tailored glove with the ostensible purposes of the decade-long, US-led intervention in Afghanistan.

However, as Greenwald shows, subsequent reporting from Associated Press' Amir Shah and Jerome Starkey of *The Times* of London pursued the story with testimony from local sources. The reporters determined that US special forces had committed the killings. Following paroxysms of denial, NATO spokespeople eventually acknowledged culpability. Greenwald concludes that "U.S. Media constantly repeats false Pentagon claims" (2010k: 4), although it may later shrug off the demonstrably false official discourse as having been a regrettable but aberrant episode. Make-believe stories that have been walked back include "Jessica Lynch's heroic firefight" as well as "Pat Tillman's death by al Qaeda"(2010k: 5). Nonetheless, Greenwald argues, the "relentless propaganda machine never seems to diminish" for the fact that dots consistently do not get connected and broad patterns of government deceit are not traced within mainstream media discourse (2010k: 5).

Greenwald's blog characterizes mainstream news accounts as laden with double standards that follow from dependence on the Us/Them dichotomy. In this vein, Greenwald confronts *New York Times* pundit Ross Douthat. In reply to threats toward the *South Park* television program for its depiction of the Prophet Muhammad as a bear, Douthat posits it as an "'example of Western institutions cowering before the threat of Islamist violence. (...) Our culture has few taboos that can't be violated (...). Except where Islam is concerned'" (2010o: 1). In answer to Douthat's assertions of brazen Islamic exceptionalism, Greenwald marshals an account from the previous month in his own newspaper. The item from the *Times* reports the cancellation of a play, *Corpus Christi* ("body of Christ" in Latin), at a Texas university. The uproar included "threatening calls and email messages" for depicting Christ as gay (2010o: 1). Greenwald shifts to the larger picture and argues that other religions' censorious bids in the US "are too numerous to chronicle" although he cites at least six pertinent examples (2010o: 2). He concludes that "the very idea that such conduct is remotely unique to Muslims is delusional, the by-product of Douthat's ongoing use of his *New York Times* column for his anti-Muslim crusades and sectarian religious promotion" (2010o: 2). For Greenwald, these causes are part and parcel to a "typically right-wing need to portray his own majoritarian group as the profoundly oppressed victim." In addendum to the slap down of Douthat, Greenwald's post also makes an assertive First Amendment move; it features the image of Muhammad that prompted the illiberal threats against *South Park*.

DemoRepublican Party Governance

Although a supporter of Obama during the 2008 election campaign, by 2010, Greenwald regularly composed scorching criticism of the record that the administration was compiling. By contrast, Kurtz says relatively little about Obama in the corpus of articles but expresses no significant criticism. However, this may indicate more about Kurtz' posture toward authority than about Obama's instantiation of it.

While Greenwald does not employ the term "sphere of consensus," it is a concept that pulls together his observations of the unfolding Obama epoch. He posits that the Democratic Party achieved electoral gains in 2006 and 2008 by differentiating itself from Bush's governance. By 2009, however, "one of the most consequential aspects of the Obama presidency" is "the conversion of numerous Bush/Cheney policies from what they once were (controversial, divisive, right-wing extremism) into what they have become (uncontroversial bipartisan consensus)" (2010p: 2).

Greenwald also frequently lambasts the Obama catchphrase of "looking forward not backward" as it effectively immunized the previous administration's "systematic war crimes, torture regime, chronic law breaking, and even human experimentation" (2010q: 1), the latter as an adjunct to the neo-conservative's torture program (Physicians for Human Rights 2010). As Greenwald observes, effective immunization not only fails to punish the perpetrators of Statist abuse, but it enables institutionalization and extension of such practices. In this vein, in early 2010, Greenwald repeatedly focused on the Obama administration's designs to assassinate US citizens "far away from any battlefield" and independent of any legal process of charges and convictions (2010l: 1). Greenwald notes that the Obama administration's legal theories had galloped further to the right than even those of Supreme Court Justice Antonin Scalia. To wit, in 2004, Scalia judged that "it was unconstitutional for the U.S. Government merely to *imprison* let alone kill American citizens as 'enemy combatants'" outside due process of law (original emphasis; 2010l: 2).

In a follow-up, Greenwald elaborates that it is "obvious" that a country reserves the right to "kill someone *on an actual battlefield during war* without due process" (original emphasis, 2010r: 1). However, in its statements, Obama's administration construes the whole world as a battlefield; and, within that unbounded sphere, the US president's powers are asserted as effectively "unlimited." This legal theory is a "core premise that spawned 8 years of Bush/Cheney radicalism (...) adopted in full by the Obama administration," that also appeared poised to act on the theory more decisively than its predecessors. Greenwald concludes that, while aggressively seizing ground from Bush/Cheney, Obama's party and supporters had been relatively silent or even supportive. Not even the previous degrees of legitimate controversy were in play to curb the extension of executive power that, once enacted, is unlikely to be rolled back by future administrations.

In each of these instances, Greenwald's legally informed posts on current issues within media discourse present an educational angle that enables readers to be more fluent on salient constitutional matters. This would appear to deliver on the high hopes for educating the public that blogging has inspired. Greenwald's method also presents a stark contrast with the mainstream media's fascination with who is winning the political tug-of-war and whose fortunes and image are on the wax or wane. Moreover, in minting these criticisms of Obama, Greenwald does not simply personalize, although the criticisms do present some personal bite. Rather, he diagnoses serious deficiencies in the US' political culture and its regression across the past generation. Greenwald writes that, at present, the notion that "the State must charge someone with a crime" and give him or her a fair trial as a condition for imprisonment "has been magically transformed into Leftist extremism" (2010g: 1). While it may be the children of the Reagan era who have accelerated the drift into Statism, Greenwald writes that, "It was also Reagan who signed the Convention against Torture in 1988" (2010g: 2). Moreover, he observes that the Reagan administration's stated policy toward terrorists was "'to get society to see them for what they are—criminals—and to use democracy's most potent tool, the rule of law, against them'" (2010g: 2).

By contrast with Reagan's avowedly right-leaning administration, in the new millennium, the rule of law in the form of due process in alleged terror cases has been marginalized to the sphere of "civil liberties extremists" (2010g: 2). Greenwald emphasizes that the US has eschewed the due process path of other nations with serious terrorism problems such as Britain, India, Indonesia, and Spain. Instead, the US has embraced the due process-cleansed practices of regimes such as Gadafi-era Libya and Saudi Arabia. In conjuring this political climate, media enablement has been a vital element; ergo, media is an unavoidable presence in Greenwald's posts.

Conclusion

The opening sections of this chapter review a body of evidence suggesting that the internet does not explode elite economic and ideological dominance. Indeed, the internet may invigorate that dominance in some degree. More broadly, the internet has been and is being conscripted for authoritarian purposes. Thus, in setting up the Kurtz-Greenwald contrast examined here, it is not enough to assert that writing from an internet platform guarantees a progressive and critical posture through techno-alchemy. As if to confirm this observation, Kurtz left the *Washington Post* for the internet platform of *The Daily Beast* in 2010 while Greenwald enacted the reverse trajectory from internet-only *Salon* to the legacy media of *The Guardian* two years later.

In lieu of technological determinist assumptions, the example that Greenwald presents may be construed as one in which the writer, the platform, and the relative autonomy from the Propaganda Model's demands coalesced into a blog that generates oppositional discourse. While it is important that a blog such as Greenwald's has achieved presence and prestige within the internet ecology, there is no technological essence that drives inexorably toward an oppositional telos over other possible outcomes. In MacKinnon's words, "many governments are realizing that their power is not necessarily doomed as a result [of new media] and have been quick to adopt techniques for manipulating the public discourse" (2012: 62). For this reason, the brutal Assad dynasty in Syria placed formerly banned Facebook back online in 2011 to better wage its campaign against rebels while simultaneously sabotaging their new media communications (MacKinnon 2012: 64-66). Moreover, the alignment between the State and private sector potentates is not loosened by the advent of new media. That nexus may be even tighter when new and inviting electronic soapboxes are owned by private corporations that are sensitive to government whims—but without statutory obligations for transparency.

If the internet is to be a liberating instrument in the politicized media battles ahead, complacent assumptions about techno-essence are a decided hindrance. If liberationist means and outcomes are to be fruits of new media, hard work and rigorous strategy will make them so. Only in that case will the stubborn prerogatives of the Propaganda Model be rolled back in favor of a better model of news discourse.

Notes

[1] Morozov also confronts the widely-circulated narrative that Twitter was driving Iran's rebellion in the summer of 2009 (2011: 1-19). He estimates that there were very few tweeters in Tehran and information about where to protest was widely available on the street without high-tech mediation. In this view, excited Westerners and the Iranian diaspora were mainly tweeting to each other about Persian tweeting.

[2] While Greenwald writes most every day, posts that were analyzed coincided with days that Kurtz had published an article in order to realize rough balance in the quantity of discourse from the two writers. On those days when a Greenwald post did not coincide with a Kurtz article, a post from the day after was used. A couple of articles could not be accessed from *Salon*'s webpage despite repeated attempts in July 2012, although Greenwald appears to have posted on those days (1 March and 8 March 2010). Whereas, the Nexis Lexis search yields precise word counts for Kurtz, Greenwald's word count was estimated by printing the corpus of his articles and then counting lines and half lines of text.

[3] As Thomas' question assumes, even a brief statistical portrait presents the unfathomable grimness of occupied Iraq. As concerns mortality, Iraq Body Count's investigators (2012) tabulate over 115,000 documented violent deaths in Iraq from 2003 to 2011, while acknowledging it is surely an undercount. Based on inferences from a sample of 12,000 civilians, Burnham, Lafta, Doocy & Roberts (2006) estimate over 650,000 excess deaths in Iraq from 2003 to 2006 in a report published in the prestigious medical journal *The Lancet*. As for the 2003 neoconservative invasion's role in attracting al Qaeda to Iraq, the Center for Strategic and International Studies' M.J. Kirdar writes, "The U.S.-led invasion and subsequent occupation provided the *raison d' être* for AQI [Al Qaeda in Iraq], fueled the influx of foreign fighters, enabled a lawless environment for profitable criminal activity, and facilitated the execution of its strategic goals" (2011:2). As concerns refugees, the United Nations High Commission on Refugees states that almost ten years after the US-led invasion, "Some one million people remain

displaced throughout the country, of whom hundreds of thousands live in dire conditions. Most are unable to return to their areas of origin because of the volatile security situation, the destruction of their homes, or lack of access to services" (United Nations High Commission on Refugees 2012: 1). On surveying evidence from prestige sources, and in contrast with Kurtz' sanctimonious finger-wagging, Thomas was on solid ground in her question to Bush.

AFTERWORD
Reboot, Retool

In the month that I am writing, Barack H. Obama secured reelection as US president by a decisive margin. The US Senate will remain in the hands of the center-left party, even though far more of its incumbent seats were tested before the electorate in the autumn of 2012. As for the House of Representatives, Republicans retained a clear majority, although Democrats attracted more votes nationwide, an advantage mooted by gerrymandered districts (Wang 2012). The 2012 election cycle was in some appreciable degree a referendum on social democratic governance that Democrats assertively underlined at their 2012 convention. The question of activist government was of further moment when Hurricane Sandy raged in the week before the balloting. As backdrop to the 2012 election cycle, a vibrant left-leaning online community is now in full bloom. Finally, incumbent news media's instinctive orientation to the mainstream and to who is perceived as winning in the political arena now clearly works in the center-left party's favor. As *New York Times* pundit Nicolas D. Kristof put it in September 2012, "When I was growing up in Oregon, it was Democrats who were typically the crazies," citing the irregular examples of segregationist George Wallace and cult leader Lyndon LaRouche. However, Kristof adds, "today's Republicans seem disproportionately untethered to reality" in ways that lose elections and hemorrhage support from pundits ever alert to mass opinion's center of gravity (2012: A27).

While these are positive developments for anyone partial to shadings of the left, they are also vulnerable and contingent. In the 2012 cycle, Republicans nominated a cardboard cutout standard bearer for their self-appointed "people's party" of the *Volk,* a plutocrats' plutocrat and millionaire *250 times over*. As if eager to confirm Democrats' talking points, Republicans also nominated several candidates for statewide office in otherwise electorally safe precincts who were, in Jimi Hendrix's words in a very different time, driven to let their freak flags fly. Candidates in Missouri and Indiana could not avoid the unforced errors of positioning as crude rape apologists spouting crackpot pieties that betrayed "movement conservativism's" gut chauvinism and authoritarianism.

To the extent that Obama's re-election was eased by his campaign's favorable intersection with demographics—youth, women, and minorities—Democrat victories may be locked in for the foreseeable future. Populous states such as California and Illinois, dominated by Republicans as recently as the 1980s, are bellwethers of the past generation's reshuffled demographics; shifts that have doomed today's "Grand Old Party" to be the plodding vehicle of angry gray men jealously guarding their stacks of money. Then again, perhaps not: A young minority candidate for president—say, Marco Rubio—could partly unpick the demographic lock in a presidential election, particularly if he or she disguises right wing radicalism as resolutely as George W. Bush's 1999-2000 campaign did.

Surely conditions are different than the 1980s, when Republicans claimed victory in 133 of 150 states (89 percent) across three presidential elections. At the same time, complacency is not in order. One interpretation of the Democratic Party's rise since the 1980s is that they have colonized the space on the political spectrum that was once occupied by centrist Republicans in the 1960s and 1970s. Simultaneously, the rest of the Republican Party has careened so far out in the political spectrum as to invite electoral spankings (although, even as a minority party, it generates havoc in the governing process via, for example, procedural abuses). Moreover, the national school house of news continues to be strongly conditioned by the Propaganda Model, as detailed in these pages. Despite the striking climatic events in 2012, with draughts and "franken-storms" all the way up to the election, Willard "Mitt" Romney's silence on the climate change issue was matched by that of Obama. In a turbulent world, US foreign policy continues to be of moment—and the sphere of consensus in foreign policy remains tight. At the same time, US audiences are still badly informed (Valentino 2012) while much the same media system—and many of the same pundits—that brayed loudly for the 2003 Iraqi invasion still ply their trade.

The paradox of the Obama era is that a cooler and hipper superstructure may in multiple registers gloss over the need for deeply inscribed change. The internet will not summon news nirvana into being by its mere existence. Similarly, the age of Obama will not arrest the problematic structuring structures of journalistic production, without retooling the news factory and making assertive moves toward rebooting journalism as the discomfort to power that its mythologies construe it to be.

REFERENCES

Ackerman, Seth (2001). "The Most Biased Name in News." *Fairness and Accuracy in Reporting/Extra!* July/August. Retrieved 16 August 2011 from: www.fair.org/index.php?page=1067.

Adegoke, Yemisi (2011). "Victims' Stories: 'They Were Rioting Outside My House.'" *The Guardian*, 7 December, p.6.

Aitkenhead, Decca (2011). "Rise Up, Rise Up." *The Guardian*, 31 December, p.13.

Allan, Stuart (2005). "Introduction." In, *Journalism: Critical Issues*, edited by Stuart Allan, pp.1–15. Maidenhead, UK: Open University Press.

Althusser, Louis (1994). "Selected Texts." In *Ideology*, edited by Terry Eagleton, pp.87–111. New York: Longman. Originally published in English 1971.

Amnesty International (2007). *Belarus—Amnesty International Report 2007*. Retrieved 21 June 2012 from: www.amnesty.org/en/region/belarus/report-2007.

Amnesty International (2006). *Israel/Lebanon: Deliberate Destruction or "Collateral Damage"?* 22 August. Retrieved 30 July 2012 from: www.amnesty.org/en/library/info/MDE18/007/2006.

Amnesty International (1997). *Belarus: Findings of Human Rights Committee Confirm Worsening Human Rights Situation*. 10 November. Retrieved 21 June 2012 from: www.amnesty.org/en/library/info/EUR49/017/1997/en.

Anderson, Benedict (1983). *Imagined Communities*. London: Verso.

Anti-Defamation League (2010). Press Release: "Glenn Beck's Remarks about Soros and the Holocaust "Offensive and Over the Top." 11 November. Retrieved 20 June 2012 from: www.adl.org/PresRele/HolNa_52/5906_52.htm.

Archer, Graeme (2011). "In Hackney, the Knives Are Being Sharpened." *The Daily Telegraph*, 10 August, p.20.

Arnot, Chris (2011). "Upward Mobility Is a Thing of the Past." *The Guardian*, 29 November, p.41.

Arons, Nicholas (1999). "Interview with Scott Ritter." 24 June. Retrieved 8 May 2012 from: www.fas.org/news/iraq/1999/07/990712-for.htm

Bagdikian, Ben H. (1992). *The Media Monopoly* (Fourth Edition). Boston: Beacon Press.

Baker, Brent, Tim Graham and Rich Noyes (2011). "Rewriting Ronald Reagan." 31 January. Media Research Center: Alexandria, VA. Retrieved 16 May 2012 from: www.mrc.org/special-reports/rewriting-ronald-reagan.

Baker, Edwin C. (2007). *Media Concentration and Democracy*. Cambridge: Cambridge University Press.

——— (1994). *Advertising and a Democratic Press*. Princeton, NJ: Princeton University Press.

Barkham, Patrick (2011). "'I Would Have Been in My Element.'" *The Guardian*, 13 August, p.33.

Barringer, Felicity (2003a). "U.N. Split Widens as Allies Dismiss Deadline on Iraq." *The New York Times*, 8 March, p.1.

——— (2003b). "Eclipsed by Events, U.N. Officials Wonder about the Past and Ponder the Future." *The New York Times*, 18 March, p.21.

Barringer, Felicity, & Michael R. Gordon (2003). "Inspector Orders Iraq to Dismantle Disputed Missiles." *The New York Times*, 22 February, p.1.

Barringer, Felicity, & David E. Sanger (2003a). "U.S. and Allies Ask the U.N. to Affirm Iraq Won't Disarm." *The New York Times*, 25 February, p.1.

——— (2003b). "U.S. Says Hussein Must Cede Power to Head Off War." *The New York Times*, 1 March, p.1.

Barstow, David (2008). "Behind TV Analysts, Pentagon's Hidden Hand." *The New York Times*, April 20. Retrieved 16 November 2010 from: www.nytimes.com/2008/04/20/us/20generals.html?pagewanted=all.

BBC News (2004). "The UK's 'Other Paper of Record.'" 19 January. Retrieved 29 March 2012 from: news.bbc.co.uk/2/hi/uk_news/3409185.stm.

Beattie, Tina, Alison Jasper, Paul D Murray, & George Ferzoco (2011). "Riots, Reasons and Reconciliation." *The Guardian*, 24 August, p.35.

Beckford, Martin, Mark Hughes, Thomas Harding & Andrew Hough (2011). "Anarchy Spreads: Riots and Looting in Manchester and Birmingham." *The Daily Telegraph*, 10 August, p.1.

Bennett, W. Lance (2001). *New: The Politics of Illusion*, Fourth Edition. New York: Addison Wesley Longman.

Benson, Rodney (2011). "Public Funding and Journalistic Independence." In *Will the Last Reporter Please Turn Out the Lights*, edited by Robert W. McChesney & Victor Pickard, pp.314–319. New York: New Press.

Benson, Rodney, & Erik Neveu (2005). "Field Theory as a Work in Progress." In *Bourdieu and the Journalistic Field*, Rodney Benson & Erik Neveu (Eds.), pp.1–25. Cambridge, UK: Polity Press.

Bird, S. Elisabeth (1990). "Storytelling on the Far Side: Journalism and the Weekly Tabloid." *Critical Studies in Mass Communication,* 7 (4), pp.377–389.

Blodget, Henry (2011). "The Golden Age of News." *Business Insider*, 19 January. Retrieved 18 December 2012 from: www.businessinsider.com/howard-kurtz-salary-2011-1.

Boehlert, Eric (2010). "Does Fox News Coverage = GOP Campaign Contribution?" *Media Matters*, 26 January. Retrieved 21 August 2011 from: mediamatters.org/columns/201001260004.

——— (2009). *Bloggers on the Bus*. New York: Free Press.

——— (2004). "Sleaze and Smear at Sinclair." *Salon*.com, 22 October. Retrieved 19 June 2012 from: www.salon.com/2004/10/22/sinclair_6/print/.

Boorstin, Daniel J. (1992). *The Image*. New York: Vintage. Originally published 1962.

Bourdieu, Pierre (2005). "The Political Field, the Social Field and the Journalistic Field." In *Bourdieu and the Journalistic Field*, edited by Rodney Benson & Erik Neveu, pp.29–47. First published 1995. Cambridge, UK: Polity Press.

——— (1999). "How Can One Be a Sports Fan?" In *The Cultural Studies Reader*, edited by Simon During, pp.427–440. New York: Routledge. Originally published 1978.

Boyd-Barrett, Oliver (2005). "Journalism, Media Conglomerates, and the Federal Communications Commission." In *Journalism: Critical Issues*, edited by Stuart Allan, pp.342-356. Maidenhead, UK: Open University Press.

——— (2004). "Judith Miller, *The New York Times*, and the Propaganda Model." *Journalism Studies*, 5(4), 435–449.

Brock, David (2002). *Blinded by the Right*. New York: Three Rivers Press.

Bumiller, Elisabeth (2003). "The President: Prohibited Missile Is the 'Tip of the Iceberg' in Iraq, Bush Says." *The New York Times*, 23 February, p.1.

Burnham, Gilbert, Riyadh Lafta, Shannon Doocy, & Les Roberts (2006). "Mortality After the 2003 Invasion of Iraq: A Cross-Sectional Cluster Sample Survey." *The Lancet*, 21 October, 368(9545), pp.1421–1428.

Burns, John F. (2003). "Iraq Shows One of Its Drones, Recalling Wright Brothers." *The New York Times*, 12 March, p.12.

Burton, Michael (2007). *In the High Court of Justice, Queen's Bench Division Administrative Court, before Mr. Justice Burton, between Stuart Dimmock and Secretary of State for Education and Skills: Judgment*, Case No.: CO / 3615 / 2007, 10 October 2007.

Businessweek (2005). "The New York Post: Profitless Paper in Relentless Pursuit." Retrieved 31 August 2011: www.businessweek.com/magazine/con-tent/05_08/b3921114_mz016.htm.

Cable News Network (2010). "CNN Programs: Anchors & Reporters: Howard Kurtz." http://edition.cnn.com/CNN/anchors_reporters/kurtz.howard.html.

Carr, Nicholas (2008). "Is Google Making Us Stupid?" *The Atlantic*, July/August. Retrieved 14 August 2102 from: www.theatlantic.com/magazine/archive/2008/07/is-google-making-us-stupid/6868/.

Carter, Bill (1991). "Few Sponsors for TV War News." *New York Times*, 7 February. Retrieved 23 August 2011 from: www.nytimes.com/1991/02/07/business/few-sponsors-for-tv-war-news.html?src=pm.

Chomsky, Noam (1988). *The Culture of Terrorism*. Boston: South End Press.

Clarke, Ramsey (1992). *The Fire This Time*. New York: Thunder's Mouth Press.

Clarke, Richard (2004). *Against All Enemies*. New York: Simon & Schuster.

CNN.com (2005). "N.Y. Times Statement Defends NSA Reporting." 16 December. Retrieved 10 February 2012 from: edition.cnn.com/2005/US/12/16/nytimes.statement/.

Columbia Journalism Review (2011). "Who Owns What? News Corporation." Retrieved 14 February 2012 from: www.cjr.org/resources/?c=newscorp.

Compaine, Benjamin (2005). *The Media Monopoly Myth*. Washington, DC: New Millennium Research Council.

——— (2004). "Domination Fantasies." *Reason*, 1 January. Retrieved 13 December 2012: reason.com/archives/2004/01/01/domination-fantasies/print.

Cooper, Mark (2011). "The Future of Journalism." In *Will the Last Reporter Please Turn Out the Lights*, edited by Robert W. McChesney & Victor Pickard, pp.320–339. New York: New Press.

Copps, Michael J. (2011). "What About the News?" In *Will the Last Reporter Please Turn Out the Lights*, edited by Robert W. McChesney & Victor Pickard, pp.289–298. New York: New Press.

Cowell, Alan (2003). "Blair, Increasingly Alone, Clings to Stance." *The New York Times*, 17 February, p.10.

Cox, Robert (2012). "The Persecution of Baltasar Garzón." *Buenos Aires Herald*, 5 February. Retrieved 7 June 2012 from:www.buenosairesherald.com/article/91897/the-persecution-of-baltazar-garzón.

Crompton, Sarah (2011). "Why Did Youngsters Who Lack for Nothing Join the Anarchy?" *The Daily Telegraph*, 18 August, p.25.

Curran, James (2012). Keynote Address: "Demystifying the Internet." Identity, Culture, and Communication Conference, April, Madrid.

——— (2011). *Media and Democracy*. London: Routledge.

Cushman, John H. Jr., & Steven R. Weisman (2003). "U.S. Says Iraq Retools Rockets For Illicit Uses", *The New York Times*, 10 March, p.1.

Daily Telegraph, The (2011a). "The Criminals who Shame our Nation." 10 August, p.21.

——— (2011b). "Watchdog Misled Media over Shooting." 13 August, p.2.

——— (2011c). "Police and Politicians Need to Act Together." 15 August, p.19.

——— (2011d). "Police Victim's Funeral." 10 September, p.18.

——— (2010). "Daily Telegraph named Paper of the Year." 24 March. Retrieved 30 March 2012 from: www.telegraph.co.uk/finance/newsbysector/mediatechnology andtelecoms/media/awards-for-telegraph/7508286/Daily-Telegraph-named-News paper-of-the-Year.html.

Davies, Caroline (2011). "Lack of Robust Police Response Helped Riots to Spread, Report Says." *The Guardian*, 29 November, p.11.

Davies, Nick (2009). *Flat Earth News*. London: Vintage.

Davies, Nick, & Amelia Hill (2011). "Missing Milly Dowler's Voicemail was Hacked by *News of the World*." *The Guardian*, 5 July, p.1.

Davis, Aeron (2010). "Politics, Journalism, and New Media." In *New Media, Old News*, edited by Natalie Fenton, pp.121–137. London: Sage.

Deep Climate (2011). "Retraction of Said, Wegman, et al, Part I." 15 May. Retrieved 12 June 2012 from: deepclimate.org/2011/05/15/retraction-of-said-wegman-et-al-2008-part-1/.

Diamond, Edwin (1993). *Behind the Times*. New York: Random House.

Diamond, Edwin, & Stephen Bates (1991). *The Spot* (Third Edition). Cambridge: Massachusetts Institute of Technology Press.

Ditmars, Hadani (2002). "Denis Halliday" (Interview). *Salon*.com, 20 March. Retrieved 28 April 2012 from: www.salon.com/2002/03/20/halliday/print.

Dodd, Vikram (2011). "Revealed: Man Whose Shooting Triggered Riots Was Not Armed." *The Guardian*, 19 November, p.1.

Downie, Leonard, Jr., & Michael Schudson (2009). "The Reconstruction of American Journalism." *Columbia Journalism Review*, 19 October. Retrieved 16 August 2011 from: www.cjr.org/reconstruction/the_reconstruction_of_american.php.

Duménil, Gérard, & Dominique Levy (2011). *The Crisis of Neoliberalism*. Cambridge, MA: Harvard University Press.

Eagleton, Terry (1991). *Ideology: An Introduction*. New York: Verso.

Eisenhower, Dwight D. (1961). Eisenhower Farewell Address (Full). Retrieved 16 December 2012 from: www.youtube.com/watch?v=CWiIYW_fBfY.

Entman, Robert M. (1989). *Democracy Without Citizens*. New York: Oxford University Press.

Ewen, Stuart (1976). *Captains of Consciousness*. New York: McGraw-Hill.

Everest, Larry (2001). "Hans von Sponeck: The Inside Story of U.S. Sanctions on Iraq" (Interview). *Revolutionary Worker*, 23 December. Retrieved 3 May 2012 from: http://rwor.org/a/v23/1130-39/1132/sponeck_iraq.htm.

Farrell, Maureen (2004). "The Clear Channel Controversy One Year On (Why Howard Stern's Woes Are Your Woes, Too)." 23 March, *BuzzFlash*.com. Retrieved 19 June 2012 from: www.buzzflash.com/farrell/04/03/far04009.html.

Fenton, Natalie (2010a). "Drowning or Waving?" In *New Media, Old News*, Natalie Fenton (Ed.), pp.3–16. London: Sage.

——— (2010b). NGOs, New Media and the Mainstream News. In *New Media, Old News*, edited by Natalie Fenton, pp.153–168. London: Sage.

Ferguson, Ben (2011). "Victims' Stories: 'No Warning. No Time to Prepare.'" *The Guardian*, 8 December, p.13.

Financial Times (2012). "FT Global 500 2012." Retrieved 24 January 2013 from: www.ft.com/intl/cms/a81f853e-ca80-11e1-89f8-00144feabdc0.pdf.

Fisher, Ian (2003a). "Top Iraqi Adviser Says He Believes War Is Inevitable." *The New York Times*, 26 January, p.1.

——— (2003b). "Scientist Gives Inspectors First Private Talk." *The New York Times*, 7 February, p.11.

Flew, Terry (2007). *Understanding Global Media*. New York: Palgrave MacMillan.

Foley, Henry C., Alan W. Scaroni, and Candice A. Yekel (2010). RA-10 Inquiry Report, 10 February. The Pennsylvania State University: University Park.

Forbes (2012). "The World's Billionaire List." March. Retrieved 22 January 2013 from: www.forbes.com/billionaires/list/.

Franklin, Bob (2008). "Newspapers: Trends and Developments." In, *Pulling Newspapers Apart*, edited by Bob Franklin, pp.1–35. London: Routledge.

——— (2005). "McJournalism: The Local press and the McDonaldization Thesis." In *Journalism: Critical Issues*, edited by Stuart Allan, pp.137–150. Maidenhead, England: Open University Press.

Freedman, Des (2010). "The Political Economy of the 'New' News Environment." In *New Media, Old News*, edited by Natalie Fenton, pp.35–50. London: Sage.

Friedman, Rose, & Milton Friedman (1980). *Free to Choose*. New York: HarcourtBrace Jovanovich.

Gainor, Dan (2010). "Fire and Ice." 3 November. Media Research Center: Alexandria, VA. Retrieved 16 May 2012 from: www.mrc.org/special-reports/uncritical-condition.

——— (2009). Comment, in "The Reconstruction of American Journalism," by Leonard Downie, Jr. & Michael Schudson. *Columbia Journalism Review*, 19 October. Retrieved 16 August 2011from: www.cjr.org/reconstruction/the_re-construction_of_american.php.

Gainor, Dan & Iris Somberg (2012). "George Soros: Godfather of the Left." 4 June. Media Research Center: Alexandria, VA. Retrieved 11 June 2012 from: www.mrc.org/special-reports/special-report-george-soros-godfather-left.

——— (2011). "George Soros: Media Mogul." 15 August. Media Research Center: Alexandria, VA. Retrieved 16 May 2012 from: www.mrc.org/special-reports/george-soros-media-mogul.

Gandy, Oscar (1982). *Beyond Agenda Setting*. Norwood, NJ: Ablex.

Gans, Herbert J. (2004). *Deciding What's News*. Evanston, IL: Northwestern University Press. First published 1979.

Garnham, Nicholas (1990). *Capitalism and Communication*. London: Sage.

Gentleman, Amelia (2001). "Parenting: 'Being Liberal Is Fine, but We Need to Be Given Back the Right to Parent.'" *The Guardian*, 11 August, p.10.

Gilligan, Andrew (2011). "The Vigilantes Tainted by Claims of Racism." *The Daily Telegraph*, 12 August, p.6.

Gillmor, Dan (2004). *We the Media*. Sebastopol, CA: O'Reilly.

Gitlin, Todd (2011). "A Surfeit of Crises." In *Will the Last Reporter Please Turn Out the Lights*, edited by Robert W. McChesney & Victor Pickard, pp.91–102. New York: New Press.

——— (1980). *The Whole World Is Watching*. Berkeley: University of California Press.

Global Research (2006). "Vatican Deplores Israel's Bombing of Lebanon." 20 July. Retrieved 30 July 2012 from: www.globalresearch.ca/PrintArticle.php?articleId=2776.

Gordon, Joy (2005). "Cool War." In *Tell Me No Lies*, edited by John Pilger, pp.541–552. London: Vintage.

Gordon, Michael R. (2003a). "The Weight of Evidence." *The New York Times*, 18 January, p.1.

——— (2003b). "To U.S., Onus Is on Hussein." *The New York Times*, 24 January, p.1.

Goss, Brian Michael (2009). "'The Left-Media's Stranglehold': Flak and Accuracy In Media Reports (2007-08)." *Journalism Studies*, 10 (4), pp.455-473.

——— (2006). "Sex Education Fantasies: Ideology and Right-Wing Science." *Southern Review*, 39 (1), pp.8–24

——— (2003). "The 2000 US Presidential Election in *Salon*.com and *The Washington Post*." *Journalism Studies*, 4 (2), pp.163–182.

——— (2002). "Deeply Concerned About the Welfare of the Iraqi People': The Sanctions Regime Against Iraq in *The New York Times* (1996 -98)." *Journalism Studies*, 3 (1), pp.83–99.

——— (2001). "'All Our Kids Get Better Jobs Tomorrow': The North American Free Trade Agreement in *The New York Times* (1993)." *Journalism and Communication Monographs*, 3 (1), pp.5–47.

——— (2000a). *Teeth-Gritting Harmony: The Ideology of Neoliberalism*. Unpublished doctoral dissertation, Institute of Communication Research, University of Illinois at Urbana-Champaign.

——— (2000b). "Hail to the Subject: The Durability (for the Moment) of Neoliberalism." *Cultural Studies: A Research Annual*, 5 (1), pp.363–393.

Graff, Garrett M. (2005). "See Howie Kurtz Run." *The Washingtonian*, 1 July. Retrieved 14 August 2012 from: www.washingtonian.com/articles/people/see-howie-kurtz-run/.

Graham, Tim (2012). "Secular Snobs." April. Media Research Center: Alexandria, VA.Retrieved 16 May 2012 from: www.mrc.org/sites/default/files/docu-ments/SecularSnobs_0.pdf.

——— (2011). "Red, White and Partisan." September. Retrieved 16 May 2012 from: www.mrc.org/sites/default/files/documents/Terrorism2011.pdf.

——— (2010). "Syrupy Minutes." September. Media Research Center: Alexandria, VA. Retrieved 16 May 2012 from: www.mrc.org/sites/default/files/documents/SyrupyMinutes.pdf.

Graham, Tim & Geoffrey Dickens (2011). "Counting the Reasons to Defund." October. Media Research Center: Alexandria, VA. Retrieved 16 May 2012 from: www.mrc.org/sites/default/files/documents/ReasonsDefund.pdf.

Greenwald, Glenn (2012). "Last Day at *Salon*." *Salon*.com, 15 August. Retrieved 16 August 2012 from: www.slaon.com/2012/08/15/last_day_at_Salon.

——— (2010a). "More Cause and Effect in Our Ever-Expanding 'War.'" *Salon*.com, 7 January. Retrieved 10 July 2012 from: www.salon.com/2010/01/07/terror-ism_16/print/.

——— (2010b). "Helen Thomas Deviates from the Terrorism Script." *Salon*.com, 9 January. Retrieved 10 July 2012 from: www.salon.com/2010/01/09/thomas_7/print/.

——— (2010c). "'Political Reporting' Means 'Royal Court Gossip.'" *Salon*.com, 11 January. Retrieved 10 July 2012 from: www.salon.com/2010/01/10/halperin_4/print/.

——— (2010d). "The Fundamental Unreliability of America's Media." *Salon*.com, 12 January. Retrieved 10 July 2012 from: www.salon.com/2010/01/10/media_254/print/.

——— (2010e). "Congress Takes a Bold Stand Against Surveillance Abuses." *Salon*.com, 18 January. Retrieved 10 July 2012 from: www.salon.com/2010/01/18/china_31/print/.

——— (2010f). "What the Supreme Court Got Right." *Salon*.com, 22 January. Retrieved 10 July 2012 from: www.salon.com/2010/01/22/citizens_united/print/.

——— (2010g). "Nostalgia for Bush/Cheney Radicalism." *Salon*.com, 31 January. Retrieved 11 July 2012 from: www.salon.com/2010/01/31/nostalgia_3/print/.

——— (2010h). "Susan Collins Spreads the Central Myth about the Constitution." *Salon*.com, 1 February. Retrieved 11 July 2012 from: www.salon.com/2010/02/01/collins_5/print/.

——— (2010i). "The Creepy Tyranny of Canada's Hate Speech Laws." *Salon*.com, 22 March. Retrieved 11 July 2012 from: www.salon.com/2010/03/22/canada_5/print/.

——— (2010j). "Mike McConnell, the *WashPost* and the Dangers of Sleazy Corporatism." *Salon*.com, 29 March. Retrieved 11 July 2012 from: www.salon.com/2010/03/29/mcconnell_3/print/.

——— (2010k). "How Americans Are Propagandized about Afghanistan." *Salon*.com, 5 April. Retrieved 11 July 2012 from: www.salon.com/2010/04/05/afghanistan_34/print/.

——— (2010l). "Confirmed: Obama Authorizes Assassination of U.S. Citizen." *Salon*.com, 7 April. Retrieved 11 July 2012 from: www.salon.com/2010/04/07/assassinations_2/print/.

——— (2010m). "The White House Seeks Out Kagan Defenders." *Salon*.com, 17 April. Retrieved 11 July 2012 from: www.salon.com/2010/04/ 7/kagan_4/print/.

——— (2010n). "The Long, Clear, Inspiring Record of Diane Wood." *Salon*.com, 19 April. Retrieved 11 July 2012 from: www.salon.com/2010/04/19/wood_3/print/.

——— (2010o). "*The New York Times*' Muslim Problem." *Salon*.com, 26 April. Retrieved 11 July 2012 from: www.salon.com/2010/04/26/douthat_4/print/.

——— (2010p). "The Absence of Debate Over War." *Salon*.com, 24 May. Retrieved 11 July 2012. from: www.salon.com/2010/05/24/wars_2/print/.

——— (2010q). "A Growing Part of the Obama Legacy." *Salon*.com, 8 June. Retrieved 11 July 2012 from: www.salon.com/2010/06/08/legacy_3/print/.

——— (2010r). "How Many Americans Are Targeted for Assassination?" *Salon*.com, 25 June. Retrieved 11 July 2012 from: www.salon.com/2010/06/25/assassination_3/print/.

——— (2010s). "New Study Documents Media's Servitude to Government." *Salon*.com, 30 June. Retrieved 11 July 2012 from: www.salon.com/2010/06/30/media_258/print/.

——— (2009). "GE's Silencing of Olbermann and MSNBC's Sleazy Use of Richard Wolffe." *Salon*.com, 1 August. Retrieved 11 December 2012 from: www.salon.com/2009/08/01/ge/.

——— (2008). "Media's Refusal to Address the NYT's Military Analyst Program Continues." *Salon*.com, 22 April. Retrieved 7 February 2012 from: www.salon.com/2008/04/22/analysts/

Greider, William (1997). *One World, Ready or Not*. New York: Simon & Schuster.

——— (1992). *Who Will Tell the People*. New York: Simon & Schuster.

Grieve, Tim (2003). "Fox: The Inside Story." *Salon*.com, 31 October. Retrieved 21 August 2012 from: dir.salon.com/story/news/feature/2003/10/31/fox/index.html.

Grueskin, Bill, Ava Seave & Lucas Graves (2011). *The Story So Far*, May 10. Retrieved 18 December 2012 from: http://www.cjr.org/the_business_of_digital_journalism/the_story_so_far_what_we_know.php.

Guardian, The (2011a). "Leading the Way in Investigative Journalism," 16 July, p.2.

——— (2011b). "Urban Riots: The Battle for the Streets." *The Guardian*, 10 August, p.28.

——— (2011c). "Urban Riots: Seven Days That Shook Britain." *The Guardian*, 13 August, p.40.

——— (2011d). "Rights and Wrong'uns: After the Riots," *The Guardian*, 29 August, p.26.

Hackett, Robert A. (2005). "Is There a Democratic Deficit in US and UK Journalism?" In *Journalism: Critical Issues*, edited by Stuart Allan, pp.85–97. Maidenhead, UK: Open University Press.

Halliday, Denis (2003). "2003 Peace Award." Ghandi Foundation, 30 January. Retrieved 28 April 2012 from: gandifoundation.org/2003/01/30/2003-peaceaward-denis-halliday-2/.

Halliday, Josh (2011). "Blank Pictures from Libyan State TV Augurs Moment of Change." *The Guardian*, 23 August, p.5.

Halliday, Josh, & Vikram Dodd (2012). "Phone Hacking: Six Arrested Under Operation Weeting." *The Guardian*, 13 March. Retrieved 24 January 2013 from: http://www.guardian.co.uk/media/2012/mar/13/phone-hacking-six-arrested.

Hallin, Daniel C. (1994). *We Keep America On Top of the World*. London: Routledge.

Hamilton, James T. (2004). *All the News That's Fit to Sell*. Princeton, NJ: Princeton University Press.

Hamilton, John (2010). "Progressive Hunter." *Media Matters*, 11 October. Retrieved 19 June 2012 from: http://mediamatters.org/research/201010110002.

Harris, John (2011). "Where Will It All End?" *The Guardian*, 19 July, pp.G2:4–G2:7.

Harris, Paul (2012). "Drone Wars and State Secrecy—How Barack Obama Became a Hardliner." *The Observer*, 3 June, p.30.

Harrison, Bennett, & Barry Bluestone (1988). *The Great U-Turn*. New York: Basic Books.

Harvey, David (2005). *A Brief History of Neoliberalism*. Oxford: Oxford University Press.

——— (1989). *The Condition of Post-Modernity*. Oxford: Blackwell.

Hattenstone, Simon (2011). "The G2 Interview: 'Just Because You're Poor Doesn't Mean That You Can't Know the Moral Difference Between Right and Wrong.'" *The Guardian*, 15 August, p.G2:6.

Hayes, Danny & Matt Guardino (2010). "Whose Views Make the News?" *Political Communication*, 27: 1, pp.59–87.

Hedges, Chris (2011). "The Disease of Objectivity." In *Will the Last Reporter Please Turn Out the Lights*, edited by Robert W. McChesney & Victor Pickard, pp.209–213. New York: New Press.

Helm, Toby (2012). "Rising Crime and Cash Cuts Open a New Front for Labour Attacks." *The Observer*, 11 March, p.8.

Herman, Edward S., & Noam Chomsky (1988). *Manufacturing Consent*. Retrieved 23 November 2012 from: www.thirdworldtraveler.com/Herman%20/Manufac_Consent_Prop_Model.html.

Hertsgaard, Mark (1989). *On Bended Knee*. New York: Schocken Books.

Hesmondalgh, David (2007). *Cultural Industries*. London: Sage.

Hickman, Leo (2012). "Environment Blog: Heartland Institute Compares Belief in Global Warming to Mass Murder." *The Guardian*, 4 May. Retrieved 13 April 2013 from: www.guardian.co.uk/environment/blog/2012/may/04/heartland-institute-global-warming-murder.

Hitchens, Christopher (2001). *The Trial of Henry Kissinger*. New York: Verso.

Hoare, Stephen (2011). "Maintaining the Thin Blue Line." *The Daily Telegraph*, 21 September, p.2.

Horton. Scott (2012). "Obama's Kill List." *Harper's*, 29 May. Retrieved 25 June 2012 from: www.harpers.org/archive/2012/05/hbc-90008639.

Hoynes, William (2002). "Why Media Mergers Matter." openDemocracy, 16 January. Retrieved 15 August 2011 from: www.opendemocracy.net/media-globalmedia ownership/article_47.jsp.

Human Rights Watch (2012). "Spain: Garzón Trial Threatens Human Rights." 13 January. Retrieved 7 June 2012 from: www.hrw.org/print/news/2012/01/13/spain-garzon-trial-threatens-human-rights.

——— (2006). *Fatal Strikes*, 2 August. Retrieved 30 July 2012 from: www.hrw.org/print/reports/2006/08/02/fatal-strikes-0.

Hutchinson, Peter (2011). "Wrestled, Punched and Mugged for My Blackberry." *The Daily Telegraph*, 10 August, p.3.

Indyk, Martin, & Kenneth M. Pollack (2003). "How Bush Can Avoid the Inspections Trap." *The New York Times*, 27 January, p.25.

Intergovernmental Panel on Climate Change (2009). "Summary Description of the IPCC Process," December. Retrieved 20 June 2012 from: https://www.ipcc-wg1.unibe.ch/statement/WGIsummary22122009.html.

Iraq Body Count (2011). "Iraqi Deaths from Violence, 2003-2011." 2 January. Retrieved 14 August 2012 from: www.iraqbodycount.org/analysis/numbers/2011/.

Iraq Survey Group (n.d.). "Iraq Survey Group Final Report: Key Findings." Retrieved 13 May 2012 from: www.globalsecurity.org/wmd/librar-y/report/2004/isg-final-report/isg-final-report_vol3_cw_key-findings.htm.

Jackson, Janine (2011). "A Better Future for Journalism Requires a Clear-Eyed View of Its Present." In *Will the Last Reporter Please Turn Out the Lights*, edited by Robert W. McChesney & Victor Pickard, pp.202–208. New York: New Press.

Jacobs, Andrew (2003). "My Week at Embed Boot Camp." *The New York Times: Sunday Magazine*, 2 March. Retrieved 20 February 2012: www.nytimes.com/2003/03/02/magazine/the-way-we-live-now-3-02-03-process-my-week-at-embed-boot-camp.html.

Jenkins, Henry (2008). *Convergence Culture.* New York: New York University Press.

Jenson, Robert (2005). "Dan Rather and the Problem with Patriotism." In *Filtering the News*, edited by Jeffrey Klaehn, pp.120–137. Montréal: Black Rose Books.

Johnson, Boris (2011). "This Is No Time to Be Squeamish." *Daily Telegraph*, 15 August, p.17.

Johnson, Daniel (2011). "In This Nightmarish Week There Have Been Countless Examples of Real Heroism, Heartwarming Kindness and Bloody-Minded Determination to Just Keep Soldiering On." *The Daily Telegraph*, 10 August, p.19.

Johnson, Mark (2011). "Inside Out: Tomorrow's Riots Can Be Prevented." *The Guardian*, 19 October, p.37.

Johnston, Philip (2011a). "The Long Retreat of Order." *The Daily Telegraph*, 10 August, p.19.

Johnston, Philip (2011b). "Silly Me, I Didn't Realise the Rioters Were Victims." *The Daily Telegraph*, 6 December, p.29.

Jolly, Rhonda (2007). Research Paper: *Media Ownership Deregulation in the United States and Australia.* Canberra: Department of Parliamentary Services.

Katz, Jon (1993). "The Media's War on Kids." *Rolling Stone*, 25 November, pp.47–49.

Kessler, Glenn (2012). "Romney's Claim That Obama Has Raised Taxes on "Millions of Americans." *Washington Post*, 13 April. Retrieved 21 August 2012 from: www.washingtonpost.com/blogs/fact-checker/post/romneys-claim-that-obama-has-raised-taxes-on-millions-of-americans/2012/04/12.

Khaleeli, Homa (2011). "Iran: Topping the World Schadenfreude League", *The Guardian*, 10 August, p.G2:2.

Kilpatrick, David (2010). *The Facebook Effect.* Chatham, UK: Virgin Books.

Kirdar, M. J. (2011). *Al Qaeda in Iraq.* June, Transnational Threats Project, Center for Strategic and International Studies, Washington, DC.

Kishtwari, Soraya (2009). "*Newsweek*'s New Look." *Editors Weblog*, 20 April. Retrieved 6 August 2009: www.editorsweblog.org/2009/04/20/newsweeks-new-look.

Kiss, Jemima (2009). "ABCe: Guardian.co.uk takes top spot." *The Guardian*, 21 May. Retrieved 14 April 2013 from: www.guardian.co.uk/me-dia/2009/may/21/abce-guardian-telegraph.

Klaehn, Jeffrey (2005). "A Critical Review and Assessment of Herman and Chomsky's 'Propaganda Model.'" In *Filtering the News*, edited by Jeffrey Klaehn, pp.1–20. Montréal: Black Rose.

Klein, Naomi (2011). "If You Rob People of the Little They Have, Expect Resistance." 18 August, *The Guardian*, p.35.

——— (2007). *The Shock Doctrine.* New York: Metropolitan Books.

Kovach, Bill, Tom Rosenstiel & Amy Mitchell (2004). "A Crisis of Confidence." In, *How Journalists See Journalists in 2004*, pp.27–32. Washington, DC: Project for Excellence in Journalism.

Kristof, Nicholas D. (2012). "It Takes One to Know One." *The New York Times*, 20 September, p.A27.
Krugman, Paul (2003). "Channels of Influence." *The New York Times*, 25 March, p.17.
Kurtz, Howard (2010a). "Signs of the Times." *The Washington Post*, 9 January, p.C4.
——— (2010b). "Post Sets Newsroom Policy for Sponsored Events." *The Washington Post*, 15 January, p.C4.
——— (2010c). "Game Change and Sourcing." *The Washington Post*, 18 January, p.C1.
——— (2010d). "Inquiring Minds Want to Know." *The Washington Post*, 22 January, p.C1.
——— (2010e). "Obama Embraces New Media, Which Piques the Old Guard." *The Washington Post*, 8 February, p.C1.
——— (2010f). "Tiger, Toyoda and the Imperfect Art of the Apology." *The Washington Post*, 1 March, p.C1.
——— (2010g). "In Lean Times, Multi-Tasking TV Reporters Aren't Just in Front of the Camera." *The Washington Post*, 8 March, p.C1.
——— (2010h). "The Beck Factor at Fox." *The Washington Post*, 15 March, p.C1.
——— (2010i). "The Press Suffering from Complications." *The Washington Post*, 22 March, p.C1.
——— (2010j). "For a Fledgling Band of Scribes, a World of News." *The Washington Post*, 25 March, p.C1.
——— (2010k). "For Authors, White House Is Hardly an Open Book." *The Washington Post*, 31 March, p.C1.
——— (2010l). "White House Slams CBS on Blog Post About Kagan's Sexuality." *The Washington Post*, 16 April, p.C1.
——— (2010m). "Nonprofit's News Gathering Pays Off." *The Washington Post*, 19 April, p.C1.
——— (2010n). "Washington Times Publisher Fired after Clashing with Editor He Hired." *The Washington Post*, 26 April, p.C1.
——— (2010o). "Obama Team Aggressive in Pursuing Media Leaks." *The Washington Post*, 30 April, p.C1.
——— (2010p). "Amid Continued Losses, Post Co. Puts Venerable *Newsweek* Magazine up for Sale." *The Washington Post*, 6 May, p.A16.
——— (2010q). "Lohan Lies Go Viral." *The Washington Post*, 7 May, p.C3.
——— (2010r). One Editor's Balancing Act on Journalism's High Wire. *The Washington Post*, 10 May, p.C1.
——— (2010s). "Man on the Move." *The Washington Post*, 24 May, p.C1.
——— (2010t). "Tape Shows Sarah Ferguson Asking for Cash." *The Washington Post*, 24 May, p.A13.
——— (2010u). "Out of Questions." *The Washington Post*, 8 June, p.C1.
——— (2010v). "Despite Gaffes, Biden Has Blossomed as Obama's Prime Spokesman." *The Washington Post*, 10 June, p.C1.
——— (2010w). "Tough Love May Have Saved 'Aunt Helen.'" *The Washington*

Post, 14 June, p.C1.

——— (2004). "Hear No Lichtblau, See No Lichtblau." *Washington Post*, 28 June, p.C1.

Landler, Mark & Alan Cowell (2003). "Powell, in Europe, Nearly Dismisses U.N.'s Iraq Report." *The New York Times*, 27 January, p.1.

Laville, Sandra (2011a). "Policing: Plastic Bullets." *The Guardian*, 10 August, p.7.

Laville, Sandra (2011b). "Tackling Violence Will Take 10 Years, Warns Gangs Expert." *The Guardian*, 20 August, p.6.

Laville, Sandra, Vikram Dodd & Helen Carter (2011). "Aftermath: After the Flames and Fury Come the Bill." *The Guardian*, 7 September, p.13.

Laville, Sandra, Vikram Dodd, Alex Hawkes, Matthew Taylor & Peter Walker (2011). "Lockdown: Police Get Tough." *The Guardian*, 10 August, p.1.

Leder, Michelle (2009). Comment, in "The Reconstruction of American Journalism," by Leonard Downie, Jr. & Michael Schudson. *Columbia Journalism Review*, 19 October. Retrieved 16 August 2011 from: www.cjr.org/reconstruction/the_reconstruction_of_american.php.

Lee-Wright, Peter (2010). "Culture Shock." In *New Media, Old News*, edited by Natalie Fenton, pp.71–86. London: Sage.

Letelier, Orlando (1976). "The Chicago Boys in Chile." *The Nation*, 28 August. Retrieved 13 December 2012 from: www.ditext.com/letelier/chicago.html.

Lewis, Justin, Michael Morgan & Andy Ruddock (2007). "Images / Issues / Impact." In *The Political Communication Reader*, edited by Ralph Negrine and James Stanyer, pp.176–180. London: Routledge.

Lewis, Justin, Rod Brookes, Nick Mosdell, & Terry Threadgold (2006). *Shoot First and Ask Questions Later*. New York: Peter Lang.

Lewis, Justin, Andrew Williams, Bob Franklin, James Thomas & Nick Mosdell (2006). *The Quality and Independence of British Journalism*. Journalism and Public Trust Project / Cardiff University: Cardiff, UK. Retrieved 23 October 2012 from: www.cardiff.ac.uk/jomec/research/researchgroups/journalismstudies/fundedprojects/qualitypress.html.

Lewis, Paul (2011a). "The Week That Shook Britain." *The Guardian*, 13 August, p.1 (Newsprint Supplement).

Lewis, Paul (2011b) Guardian/LSE Investigation: "There Is an Urgent Need for Rigorous Social Research." *The Guardian*, 6 September, p.11.

Lewis, Paul, & Vikram Dodd (2011). "Policing: Senior Police Face Up to Claim They Lost Control after Gaining Enemies." *The Guardian*, 7 December, p.8.

Lewis, Paul, Tim Newburn, Matthew Taylor & James Ball (2011). "Blame the Police: Why the Rioters Say They Took Part." *The Guardian*, 5 December, p.1.

Lieberman, Jon (2004). "Why I Stood Up to Sinclair." *Broadcasting & Cable*, 24 October. Retrieved 17 August 2011 from: http://www.broadcastingcable.com/article/155031-Why_I_Stood_Up_to_Sinclair.php.

Lindner, Andrew M. (2008). "Controlling the Media in Iraq." *Contexts*, Spring. Retrieved 18 February 2012 from: contexts.org/articles/spring-2008/controlling-the-media-in-iraq/.

Lovett, Ian, & Will Carless (2012). "Iraqi Immigrants in California Town Fear a Hate Crime in a Woman's Killing." *The New York Times*, 28 March, p.12.

Lynas, Mark & George Monbiot (2011). "*The Spectator* Runs False Sea-Level Claims on its Cover." *The Guardian*, 2 December. Retrieved 29 May 2012 from: www.guardian.co.uk/environment/georgemonbiot/2011/dec/02/spectator-sea-level-claims.

MacKinnon, Rebecca (2012). *Consent of the Networked*. New York: Basic Books.

MacFarquhar, Neil (2003a). "Iraq Says It Will Reply to U.N. Arms Queries." *The New York Times*, 10 January, p.11.

——— (2003b). "Iraqi Says Arms Destruction Will Cease if U.S. Attacks", *The New York Times*, 3 March, p.8.

Malik, Shiv (2011). "Muslims: For Some, Riots Were Reaction to Islamophobia." *The Guardian*, 9 December, p.22.

Mann, Michael E. (2012). *The Hockey Stick and the Climate Wars*. New York: Columbia University Press.

Massing, Michael (2004). "Now They Tell Us." *New York Times Review of Books*, 29 January. Retrieved 13 May 2012 from: www.nybooks.com/articles/archives/2004/feb/26/now-they-tell- us/?pagination=false.

McChesney, Robert W. (2004). "The Escalating War Against Corporate Media." *Monthly Review*, March. Retrieved 21 August 2011 from: monthlyreview.org/2004/03/01/the-escalating-war-against-corporate-media.

Media Matters for America (2005). "MRC Studies That 'Prove' Media's 'Liberal Bias' Collapse under Scrutiny." Retrieved 29 May 2005 from: mediamatters.org/print/research/200505110005.

——— (2004). "The World According to Coulter." 4 October. Retrieved 3 May 2012 from: mediamatters.org/research/200410040009.

Media Research Center (n.d.). "About Us." Retrieved 16 May 2012 from: www.mrc.org/static/about-us.

——— (2008). *Annual Report*. Media Research Center: Alexandria, VA.

Media Research Center/Culture and Media Institute (2012). "Baptism by Fire." 7 March. Media Research Center: Alexandria, VA. Retrieved 16 May 2012 from: www.mrc.org/special-reports/baptism-fire.

Meech, Peter (2008). "Advertising." In *Pulling Newspapers Apart*, edited by Bob Franklin, pp.235–243. London: Routledge.

Meyer, Jane (2010). "Covert Operations". *The New Yorker*, 30 August. Retrieved 20 June 2012 from: www.newyorker.com/reporting/2010/08/30/100830fa_fact_mayer.

Middleton, Christopher (2011). "The Proof that Funding Charity Projects Is Making a Difference." *The Daily Telegraph*, 10 September, p.15.

Miller, Judith (1997a). "Belarus Fines Soros Foundation $3 Million in Apparent Crackdown." *New York Times*, 2 May, p.3.

——— (1997b). "A Promoter of Democracy Angers the Authoritarians." *New York Times*, 12 July, p.1.

Milne, Seamus (2012). "America's Murderous Drone Campaign Is Fuelling Terror." *The Guardian*, 29 May, p.34.

——— (2011). "These Riots Reflect a Society Run on Greed and Looting." *The Guardian*, 11 August, p.31.

Moderate Voice, The (2012). "Interview with Colonel Morris Davis." Retrieved 3 June 2012 from: themoderatevoice.com/136355/interview-with-col-morris-davis/.

Moeller, Susan D. (2004). *Media Coverage of Weapons of Mass Destruction*. College Park; MD: Center for International and Security Studies at Maryland.

Morozov, Evgeny (2011). *The Net Delusion*. London: Allen Lane.

MSNBC (2011). Joe Scarborough. Retrieved 18 February 2012 from: www.msnbc.msn.com/id/3080460/ns/msnbc_tv-morning_joe/t/joe-scarborough/.

Mueller, John, & Karl Mueller (1999). "Sanctions of Mass Destruction." *Foreign Affairs*, May/June, pp.43–65.

Muir, Hugh & Yemisi Adegoke (2011). "Ethnicity: These Were Not Race Riots, But for Many, Race Was Still an Issue." *The Guardian*, 9 December, p.22.

Muir, Hugh & Diane Taylor (2011). "Mark Duggan Funeral." *The Guardian*, 10 September, p.14.

Murdoch, James (2009). "MacTaggert Lecture: The Absence of Trust." 28 August. Edinburgh International Television Festival. Retrieved 29 August 201 from: image.guardian.co.uk/syfiles/Media/documents/2009/08/28/JamesMurdochMac TaggartLecture.pdf.

Murphy, Joe (2011). "Vince Cable: I Declared War on Murdoch...Now Everyone Agrees with Me." *London Evening Standard*, 15 July. Retrieved 10 February 2012 from: www.thisislondon.co.uk/standard/politics/article-23970384-cable-i-declared-war-on-murdoch-and-8201-now-everyone-agrees-with-me.do.

Nagourney, Adam, & Janet Elder (2003). "More Americans Now Faulting U.N. on Iraq, Poll Finds." *The New York Times*, 11 March, p.1.

Nature (2011). "Editorial: Copy and Paste." 26 May, pp.419–420.

Neiger, Motti, Eyal Zanberg & Oren Meyers (2010). "Communicating Critique: Toward a Conceptualization of Journalistic Criticism." *Communication, Culture & Critique*, 3 (3), pp.377–395.

Nelson, Fraser (2011). "We are in Danger of Forgetting the Riots—and Ignoring Their Lessons." *The Daily Telegraph*, 25 November, p.32.

New York Times, The (2011). "Conrad M. Black." 28 June. Retrieved 7 May 2012 from: topics.nytimes.com/top/reference/timestopics/people/b/conrad_m_black/index.html.

——— (2010). "Editorial: We Can't Tell You." 4 April, p.WK8.

——— (2003a). "Report to U.N. by the Chief Inspector for Biological and Chemical Arms." 28 January, p.10.

——— (2003b). "An Improvised March to War." 2 February, p.14.

——— (2003c). "Back to the United Nations." 13 February, p.40.

——— (2003d). "Saying No to War." 9 March, p.12.

Noam, Eli M. (2009). *Media Concentration in America*. Oxford: Oxford University Press.

Noyes, Rich & Geoffrey Dickens (2011). "Still Thrilled by Obama," 15 November. Media Research Center: Alexandria, VA. Retrieved 16 May 2012 from: www.mrc.org/special-reports/still-thrilled-obama.

O'Connell, Mary Ellen (2002). "Compelling Saddam, Restraining Bush." *Jurist*, 21 November. Retrieved 27 April 2012 from: jurist.law.pitt.edu/forum/forumnew 73.php.

Odone, Cristina (2011). "Immigrants Love This Country More Than We Do." *The Daily Telegraph*, 11 August, p.20.

Oreskes, Naomi & Erik M. Conway (2010). *Merchants of Doubt*. New York: Bloomsbury Press.

Packard, Vance (2007). *The Hidden Persuaders*. New York: Ig Publishing.

Patel, Raj (2007). *Stuffed and Starved*. London: Portobello Books.

Payne, Verity (2011). "Rising Incredulity at *The Spectator*'s Use of Dubious Sea Level Claims." *The Carbon Blog*, 2 December. Retrieved 29 May 2012 from: www.carbonbrief.org/blog/2011/12/rising-incredulity-at-the-spectator%E2%80 %99s-use-of-dubious-sea-level-claims.

Pearson, Allison (2011). "Thou Shalt Happily Riot to Your Heart's Content." *The Daily Telegraph* 8 December, p.33.

Pedro, Joan (2011). "The Propaganda Model in the Early 21st Century." *International Journal of Communication*, 5, pp.1865–1905.

Physicians for Human Rights (2010). *Experiments in Torture* (White Paper). Cambridge, MA: Physicians for Human Rights.

Philbin, Matt (2012). "Off Camera." 24 February. Media Research Center: Alexandria, VA. Retrieved 16 May 2012 from: www.mrc.org/special-reports/camera-networks-ignore-liberal-hollywood%E2%80%99s-influence-dc.

Phillips, Angela (2010). "Old Sources, New Bottles." In *New Media, Old News*, edited by Natalie Fenton, pp.87–101. London: Sage.

Phillips, Angela, Nick Couldry & Des Freedman (2010). "An Ethical Deficit?" In *New Media, Old News*, edited by Natalie Fenton, pp.51–67. London: Sage.

Pike, Drummond (2010). "Why Does the Right Hate Soros?" *Politico*, 29 October. Retrieved 9 July 2012 from: dyn.politico.com/printstory.cfm?uuid=F4EAE622-F21F-D0D0-838EF3CE24F3D480.

Pilger, John (2005). *Tell Me No Lies*. London: Vintage.

——— (2000). *Paying the Price: Killing the Children of Iraq*. Network DVD, UK, 75 minutes.

Pizzo, Stephen, Mary Fricker & Paul Muolo (1989). *Inside Job*. New York: McGraw-Hill.

Pollack, Kenneth M. (2003). "A Last Chance to Stop Iraq." *The New York Times*, 21 February, p.27.

Ponticelli, Jacopo & Han-Joachim Voth (2011). "Cuts and Riots: They're Linked." *The Guardian*, 17 August, p.26.

Popular Republic of China, the Federation of Russia, and France (2002). "Irak." 8 November. Retrieved 27 April 2012 from: www.un.int/france/docu-ments_anglais/021108_cs_france_irak_2.htm.

Postman, Neil (1986). *Amusing Ourselves to Death.* New York: Penguin Books.

Prasad, Raekha (2011). "Rebels with a Cause?" *The Guardian,* 5 December, p.2.

Preston, Julia (2003a). "U.N. Study Sees 500,000 Iraqis Facing Injury In Case of War." *The New York Times*, 8 January, p.11.

——— (2003b). "U.N. Inspectors Criticize Iraqis Over Arms List." *The New York Times*, 10 January, p.1.

——— (2003c). "Annan Says Talk of War With Iraq Is Premature, but Warns Baghdad It Must Disarm." *The New York Times*, 15 January, p.8.

——— (2003d). "U.N. and U.S. Say Key Data Are Still Missing from Iraqis." *The New York Times*, 25 January, p.8.

——— (2003e). "U.N. Inspector Says Iraq Falls Short on Cooperation." *The New York Times*, 28 January, p.1.

——— (2003f). "U.N. Estimates Rebuilding Iraq Will Cost $30 Billion." *The New York Times*, 31 January, p.10.

Preston, Julia & Steven R. Weisman (2003a). "Powell to Charge Iraq Is Shifting Its Illegal Arms to Foil Inspectors," *The New York Times*, 5 February, p.1.

——— (2003b). "France Offering Plan to Expand Iraq Arms Hunt." *The New York Times*, 12 February, p.1.

Price, Lance (2006). "Rupert Murdoch Is Effectively a Member of Blair's Cabinet." 1 July, *The Guardian.* Retrieved 11 February 2012 from: www.guardian.co.uk/commentisfree/2006/jul/01/comment.rupertmurdoch.

Project for Excellence in Journalism (2003). *Does Ownership Matter in Local News: A Five Year Study of Ownership and Quality.* Washington, DC: Project for Excellence in Journalism.

Reiff, David (1993). "Multiculturalism's Silent Partner." *Harper's Magazine*, August. Retrieved 14 February 2012 from: www.harpers.org/archive/1993/08/00-01369.

Ricks, Thomas E. (2007). *Fiasco.* New York: Penguin Books.

Risen, James (2003). "Bush's Speech Puts New Focus on State of Intelligence Data." *The New York Times*, 29 January, p.11.

Redden, Joanna & Tamara Witschge (2010). "A New News Order." In *New Media, Old News*, edited by Natalie Fenton, pp.171–186. London: Sage.

Reuters (2011). "Factbox: Arrests in News International Phone-Hacking Probe." Retrieved 29 August 2011 from: /uk.reuters.com/article/2011/08/18/uk-newscorp-hacking-arrests-idUKTRE77H30F20110818.

Rogak, Lisa (2005). "On His Years as a Singer/Songwriter in Los Angeles." *The Official Website of The Man Behind* The da Vinci Code*: An Unauthorized Biography of Dan Brown.* Retrieved 13 February 2012 from: www.danbrownbio.com/excerpt2.html.

Rothstein, Betsy (2011). "Fox News Stays Silent on Hacking as Rivals Scent Blood." *The Guardian*, 18 July. Retrieved 29 August from: www.guardian.co.uk/comment isfree/cifamerica/2011/jul/18/fox-news-rupert-murdoch-sunday-talk-shows.

Rushe, Dominic (2012). "Iraqi Woman Murdered in US by Killer Who Left Note That Said: 'You're a Terrorist.'"*The Guardian*, 26 March, p.3.

Sabbagh, Dan, & Amelia Hill (2012). "Lies Evasions, Cover-Ups: How Murdoch Firm Hid Hacking Trail." *The Guardian*, 20 January, p.1.

Said, Edward W. (1979). *Orientalism*. New York: Vintage Books.

Salih, Barham A. (2003). Give Us a Chance to Build a Democratic Iraq. *The New York Times*, 5 February, p.A27.

Sands, Philippe (2010). "Philippe Sands Chilcot Inquiry Submissions." *Scribd*, 4 October. Retrieved 28 April 2012 from: www.scribd.com/doc/38686920/Phillip e-Sands-Chilcot-inquiry-submissions.

Sanger, David E. (2003a). "Bush Officials Debate Release Of Iraq Secrets," *The New York Times*, 30 January, p.1.

——— (2003b). "A Decision Made, and Its Consequences," *The New York Times*, 17 March, p.12.

Sanger, David E. & Warren Hoge (2003). "Bush and Two Allies Seem Set for War to Depose Hussein", *The New York Times*, 17 March, p.1.

Sanger, David E., & Thom Shanker (2003). "U.S. Aides Dismiss Moves by Baghdad But Feel Pressure," *The New York Times*, 4 March, p.1.

Santora, Marc (2003). "Not Enough Supplies or Money, Relief Groups Say," *The New York Times*, 12 March, p.16.

Scatamburlo-D'Annibale, Valerie (2005). "In Sync." In, *Filtering the News*, edited by Jeffrey Klaehn, pp.21–62. Montréal: Black Rose.

Schell, Jonathan (2010). "The Protocols of Rupert Murdoch." *Project Syndicate*, 20 December. Retrieved 20 June 2012 from: http://www.project-syndicate.org/print/the-protocols-of-rupert-murdoch.

Schofield, Andrew (2011). "Either We Grasp the Chance Now to Prevent Future Riots or the Mob Will Think It Is Invincible." *The Daily Telegraph*, 12 August, p.21.

Schudson, Michael (2011). *The Sociology of News*, Second Edition. New York: W. W. Norton.

Seymour, Julia A. (2012a). "Ignoring Science, 97% of Stories Hype BPA as Health Threat." 8 February. Media Research Center: Alexandria, VA. Retrieved 16 May 2012 from: http://www.mrc.org/special-reports/ignoring-science-97-stories-hype-bpa-health-threat.

Seymour, Julia A. (2012b). "Lefty Group Uses Pink to Camouflage Green Activism, Attack Business." 8 March. Media Research Center: Alexandria, VA. Retrieved 16 May 2012 from: http://www.mrc.org/special-reports/lefty-group-uses-pink-camouflage-green-activism-attack-business.

——— (2011). "Science Fiction." 23 May. Media Research Center: Alexandria, VA. Retrieved 16 May 2012 from: http://www.mrc.org/special-reports/science-fiction.

——— (2010). "Obama the Tax Cutter." 20 October. Media Research Center: Alexandria, VA. Retrieved 16 May 2012 from: www.mrc.org/special-reports/obama-tax-cutter-network-fairy-tale.

Shirky, Clay (2008). *Here Comes Everybody*. London: Penguin.

Simon, David (2011). "Build the Wall." In *Will the Last Reporter Please Turn Out the Lights*, edited by Robert W. McChesney & Victor Pickard, pp.45–54. New York: New Press.

Sims, Brian, Jack Zylman, Peter Hickey, & Danielle LeClair (2000). *Iraq Trip Report*. Washington, DC: Congressional Staff Offices of Danny K. Davis, Sam Gejdenson, Earl Hilliard, and Bernard Sanders.

Slaughter, Anne-Marie (2003). "Good Reasons for Going Around the U.N." *The New York Times*, 18 March, p.33.

Smith, Adam (1904). *An Inquiry into the Nature and Causes of the Wealth of Nations,* Book I: Chapter 5. Edwin Cannan, ed. Library of Economics and Liberty. Retrieved February 14, 2012 from: www.econlib.org/library/Smith/smWN2.html. Originally published 1776.

Smith, Terry (2011). "A Family Business." *The Guardian*. 19 July, p.24.

Soley, Lawrence (1995). *Leasing the Ivory Tower*. Boston: South End Press.

Sonwalkar, Prasun (2005). "Banal Journalism." In, *Journalism: Critical Issues*, edited by Stuart Allan, pp.261–273. Maidenhead, UK: Open University Press.

Soros, George (n.d.). "Open Society." In *Opening the Soviet System*. Retrieved 12 June 2012 from: www.osi.hu/oss/ch2e.html.

Sourcewatch (n.d.). "Sinclair Broadcast Group." Retrieved 16 August from: www.sourcewatch.org/index.php?title=Sinclair_Broadcast_Group.

Starr, Paul (2011). "Goodbye to the Age of Newspapers (Hello to an Era of Corruption." In *Will the Last Reporter Please Turn Out the Lights*, edited by Robert W. McChesney & Victor Pickard, pp.18–37. New York: New Press.

Stecklow, Steve, Aaron O. Patrick, Martin Peers & Andrew Higgins (2007). "In Murdoch's Career, A Hand on the News." *Wall Street Journal*, 5 June, p.A1.

Stern, Nicholas (2006). *Stern Review: The Economics of Climate Change: Executive Summary* (Long Version). Retrieved 27 June 2012 from: www.hm-treasury.gov.uk/d/Executive_Summary.pdf.

Sterne, Jonathan (2012). "What If Passivity Is the New Passivity?" *FlowTV*, 9 April. Retrieved 23 July 2012 from: flowtv.org/2012/04/the-new-passivity/.

Stevenson III, Adlai E. (2003). "Different Man, Different Moment." *The New York Times*, 7 February, p.25.

Stevenson, Richard W. (2003). "France and China Rebuff Bush on Support for Early Iraq War." *The New York Times*, 8 February, p.1.

Stevenson, Richard W. & Julia Preston (2003). "Bush Meets Blair Amid Signs of Split on U.N. War Role." *The New York Times*, 1 February, p.1.

Stevenson, Richard W. & David E. Sanger (2003). "U.S. Resisting Calls For a Second U.N. Vote On a War With Iraq." *The New York Times*, 16 January, p.1.

Stolberg, Sheryl Gay (2003). "Democrats Try to Turn Debate Back to Home." *The New York Times*, 15 February, p.14.

Stross, Randall (2008). *Planet Google*. London: Atlantic Books.

Sullivan, Stacy (2008). "Confessions of a Former Guant." *Salon*.com, 23 October. Retrieved 18 June 2012 from: http://www.salon.com/2008/10/23/vandeveld/.

Suskind, Ron (2004). "Without a Doubt." *The New York Times Magazine*, 17 October, p.44+.

Swaine, Jon (2011). "Supercop's Battle Order for Tackling Britain's Street Gangs." *The Daily Telegraph*, 13 August, p.4.

Tagliabue, John (2003). "European Nations Fall Short Of Consensus on Iraq Report." *The New York Times*, 28 January, p.12.

Taguba, Antonio (2008) Preface to *Broken Laws, Broken Lives*. Retrieved 20 June 2012 from: brokenlives.info/?page_id_23.

Tapper, Jake (2001). *Down and Dirty*. Boston: Little, Brown.

Topping, Alexandra, & Fiona Bawdon (2011). "Looting: Shopping for Free: How the Consumer Generation Took What It Wanted." *The Guardian*, 6 December, p.4.

Topping, Alexandra, Rebekah Diski & Helen Clifton (2011). "Gender: 'We Just Went In and Petrol-Bombed It.'" *The Guardian*, 10 December, p.16.

Townsend, Peter (1971). "Won't Get Fooled Again." *Who's Next*, performed by The Who, Polydor/MCA.

Travis, Alan (2011). "Gangs Did Not Play Central Role in Riots, Inquiry Finds." *The Guardian*, 25 October, p.14.

Travis, Alan, James Ball & Fiona Bawden (2011). "Investigating England's Summer of Disorder." *The Guardian*, 16 September, p.25.

Tremlett, Giles (2011). "Telling True Stories: Tales from the Trenches of a Foreign Correspondent in Spain." Guest lecture, Saint Louis University, Madrid, 12 April.

Tremlett, Giles & Sophie Arie (2003). "Spain/Italy: Aznar Faces 91 Percent Opposition to the War." *The Guardian*, 29 March, p.9.

Turley, Jonathan (2011). "Baltasar Garzón Receives Human Rights Award and Criticizes Obama Administration for Violations of International Law." 15 May. Retrieved 6 June 2012 from: jonathanturley.org/2011/05/15/baltasar-garzon-received-human-rights-award-and-criticizes-obama-administration-for-violations-of-international-law/.

United Nations (1999). "Report of the Second Panel Established Pursuant to the Note by the President of the Security Council of 30 January 1999 (S/1999/100), Concerning the Current Humanitarian Situation in Iraq." Retrieved 28 April 2012 from: www.casi.org.uk/info/panelrep.html.

United Nations Development Program (n.d.). "Human Development Reports." Retrieved 21 August from: hdr.undp.org/en/statistics/.

United Nations High Commission on Refugees (2012). 2012 UNHCR Country Operations Profile—Iraq. Retrieved 15 August 2012: www.unhcr.org/pages/49e486426.html.

University of East Anglia (2010). University of East Anglia's Response, 2 September. Retrieved 29 May 2012 from: www.uea.ac.uk/mac/comm/media/press/CRUstatements/independentreviews/UEAreviewresponse.

Valentino, Benjamin (2012). Datasets: Survey on Foreign Policy and American Overseas Commitments. Retrieved 12 November 2012 from: www.dartmouth.edu/~benv/data.html.

Van Dijk, Teun A. (n.d.). *Ideology and Discourse.* Retrieved 23 February 2004 from: www.discourse-in-society.org/ideo-dis2.htm.

Van Natta, Don, Jr., & David Johnston (2003). "A Terror Lieutenant with a Deadly Past", *The New York Times*, 10 February, p.1.

Wang, Sam (2012). "The House—New, with Less Democracy!" *Princeton Election Consortium*, 9 November. Retrieved 23 November 2012 from: election.princeton.edu/2012/11/09/the-new-house-with-less-democracy/.

Wasko, Janet (2001). *Understanding Disney*. Cambridge, UK: Polity Press.

Waters, Clay (2010). "Supremely Slanted." 29 September. Media Research Center: Alexandria, VA. Retrieved 16 May 2012 from: www.mrc.org/special-reports/uncritical-condition.

Watt, Nicholas, Patrick Wintour & Vikram Dodd (2011). "Politics: Johnson Sparks Debate over Reduction in Police Numbers." *The Guardian*, 11 August, p.8.

Weisman, Steven R. (2003a). "U.S. Accelerates Its Efforts to Build a Case Against Iraq." *The New York Times*, 19 January, p.12.

——— (2003b). "U.S. Set to Demand That Allies Agree Iraq Is Defying U.N." *The New York Times*, 23 January, p.1.

——— (2003c). "Patience Gone, Powell Adopts Hawkish Tone", *The New York Times*, 28 January, p.1.

——— (2003d). "Powell, in U.N. Speech, Presents Case to Show Iraq Has Not Disarmed." *The New York Times*, 6 February, p.1.

——— (2003e). "To White House, Inspector Is Now More a Dead End Than a Guidepost." *The New York Times*, 2 March, p.13.

——— (2003f). "Bush, in Rebuff to Partners, Freezes Mideast Peace Plan." *The New York Times*, 10 March, p.8.

Welch, Matt (2011). "When Losers Write History." In *Will the Last Reporter Please Turn Out the Lights*, edited by Robert W. McChesney & Victor Pickard, pp.214–222. New York: New Press.

——— (2003). "Open Season on 'Open Society.'" *Reason*, 8 December. Retrieved 29 May 2012 from: reason.com/archives/2003/12/08/open-season-on-open-society.

Wexler, Celia Viggo (2005). *The Fallout From the Telecommunications Act of 1996.* Washington: Common Cause Education Fund.

Wilkerson, Lawrence (2009). Some Truths about Guantánamo Bay. *Washington Note*, 17 March. Retrieved 20 June 2012 from: www.thewashingtonnote.com/archives/2009/03/some_truths_abo/.

Wilkinson, Richard (2005). *The Impact of Inequality*. New York: New Press.

Williams, Rachel (2011). "Interview Emeka Egbuonu: Positive Ambition." *The Guardian*, 21 September, p.37.

Wines, Michael (2003). "Russia, U.N. Swing Vote, Plays a Waiting Game." *The New York Times*, 13 February, p.19.

Winston, Brian (1995) "How Are Media Born and Developed?" In *Questioning the Media*, edited by John Downing, Ali Mohammadi, and Annabelle Sreberny-Mohammadi, pp.54–74. Thousand Oaks, CA: Sage.

Wintour, Patrick & Paul Lewis (2011). "Celebrity Culture Fuelled the Riots, Says Duncan Smith." *The Guardian*, 10 December, p.1.

Woodworth, Philip L. (2005). "Have There Been Large Recent Sea Level Changes in the Maldives Islands?" *Global and Planetary Change*. Retrieved 28 May 2012 from: www.sciencedirect.com/science/article/pii/S0921818103001085.

Wittstock, Melinda (2000). "Cousin John's Calls Tipped Election Tally." *The Observer*, 19 November. Retrieved 14 February 2012 from: www.guardian.co.uk/world/2000/nov/19/uselections2000.usa2.

Wolfe, Alan (2000). "Hobbled from the Start." *Salon*.com, 15 December. Retrieved 7 May 2012 from: http://www.salon.com/2000/12/15/trust_4/.

Zaitchik, Alexander (2009). "The Making of Glenn Beck." *Salon*.com, 21 September. Retrieved 17 August 2011 from: www.salon.com/news/feature/2009/09/21/glenn_beck.

Zeller, Tom (2003). "How to Win Friends and Influence Small Countries." *The New York Times*, 16 March, p.3.

INDEX

C

D

E

F

G

H

M

N

W

Y

Z

Intersections in Communications and Culture

Global Approaches and Transdisciplinary Perspectives

General Editors: Cameron McCarthy & Angharad N. Valdivia

An Institute of Communications Research, University of Illinois Commemorative Series

This series aims to publish a range of new critical scholarship that seeks to engage and transcend the disciplinary isolationism and genre confinement that now characterizes so much of contemporary research in communication studies and related fields. The editors are particularly interested in manuscripts that address the broad intersections, movement, and hybrid trajectories that currently define the encounters between human groups in modern institutions and societies and the way these dynamic intersections are coded and represented in contemporary popular cultural forms and in the organization of knowledge. Works that emphasize methodological nuance, texture and dialogue across traditions and disciplines (communications, feminist studies, area and ethnic studies, arts, humanities, sciences, education, philosophy, etc.) and that engage the dynamics of variation, diversity and discontinuity in the local and international settings are strongly encouraged.

LIST OF TOPICS

- Multidisciplinary Media Studies
- Cultural Studies
- Gender, Race, & Class
- Postcolonialism
- Globalization
- Diaspora Studies
- Border Studies
- Popular Culture
- Art & Representation
- Body Politics
- Governing Practices
- Histories of the Present
- Health (Policy) Studies
- Space and Identity
- (Im)migration
- Global Ethnographies
- Public Intellectuals
- World Music
- Virtual Identity Studies
- Queer Theory
- Critical Multiculturalism

Manuscripts should be sent to:

Cameron McCarthy OR **Angharad N. Valdivia**
Institute of Communications Research
University of Illinois at Urbana-Champaign
222B Armory Bldg., 555 E. Armory Avenue
Champaign, IL 61820